ASSOCIATION OF RESEARCH LIBRARIES
WASHINGTON, D.C.
2006

ARL STATISTICS 2004-05

A COMPILATION OF STATISTICS
FROM THE ONE HUNDRED AND TWENTY-THREE MEMBERS OF THE
ASSOCIATION OF RESEARCH LIBRARIES

Compiled and Edited by

MARTHA KYRILLIDOU
MARK YOUNG

ARL Statistics is published annually by

Association of Research Libraries
21 Dupont Circle, NW, Suite 800
Washington, D.C. 20036
Telephone: (202) 296-2296
FAX: (202) 872-0884
email: pubs@arl.org

The *ARL Statistics* datafiles,
with accompanying documentation, are available at http://www.arl.org/stats/

The tables presented in this publication are not indicative of performance and outcomes and should not be used as measures of library quality. In comparing any individual library to ARL medians or to other ARL members, one must be careful to make such comparisons within the context of differing institutional and local goals and characteristics.

ISBN 1-59407-742-8
ISSN 0147-2135

The paper used in this publication meets the minimum requirements of the American National Standard for Information Science and National Information Standards Organization standard—Permanence of Paper for Publications and Documents in Libraries and Archives, ANSI/NISO Z39.48-1992(R1997).

Printed in the United States of America

TABLE OF CONTENTS

GRAPHS

TABLES

RESEARCH LIBRARY TRENDS

ARL Statistics 2004-05 is the latest in a series of annual publications that describe collections, staffing, expenditures, and service activities for the 123 members of the Association of Research Libraries (ARL). Of these, 113 are university libraries; the remaining 10 are public, governmental, and nonprofit research libraries. ARL member libraries are the largest research libraries in North America, representing 16 Canadian and 107 U.S. research institutions. The academic libraries, which comprise about 92% of the membership, include 14 Canadian and 99 U.S. libraries.

Statistics have been collected and published annually for the members of the Association of Research Libraries since 1961-62, and the data are available through an interactive Web interface. Prior to 1961-62, annual statistics for university libraries were collected by James Gerould, first at the University of Minnesota and later at Princeton University.[1] These data, covering the years 1907-08 through 1961-62, are now called the Gerould statistics.[2] The whole data series from 1908, which is available on the ARL FTP server,[3] represents the oldest and most comprehensive continuing library statistical series in North America.

ARL libraries are a relatively small subset of libraries in North America, but they do account for a large portion of academic library resources in terms of assets, budgets, and the number of users they serve. The total library expenditures of all 123 member libraries in 2004-05 was almost $3.6 billion; from that, roughly $2.68 billion was spent by the 113 university libraries and more than $900 million by the nonuniversity libraries. The pie charts below show how the two types of libraries divide these expenditures differently.

University Libraries

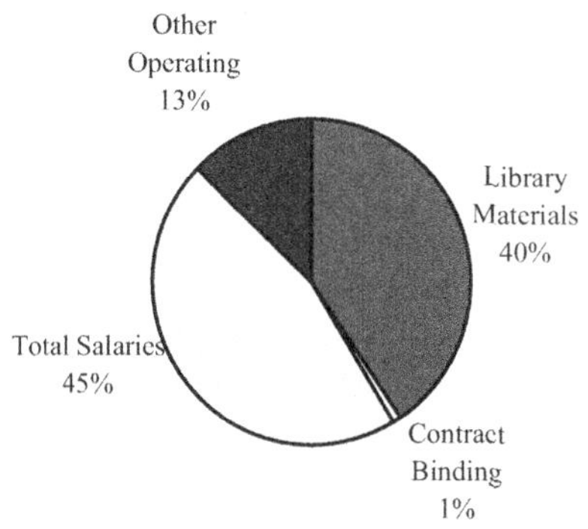

Nonuniversity Libraries

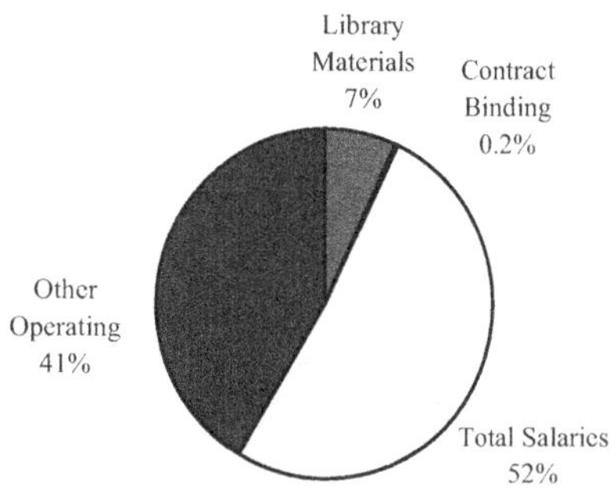

---

[1] Kendon L. Stubbs and Robert E. Molyneux, *Research Library Statistics 1907-08 through 1987-88* (Washington, DC: ARL, 1990).
[2] Robert E. Molyneux, *The Gerould Statistics 1907/08 – 1961/62*. (Washington, DC: ARL, 1986), http://fisher.lib.virginia.edu/gerould/.
[3] http://www.arl.org/stats/arlstat/mrstat.html.

EVOLUTION OF DEFINITIONS: A MOVING TARGET

Definitions of the categories used in the *ARL Statistics* are based on the *Library Statistics*, ANSI/NISO Z39.7-1995 (Bethesda, MD: NISO Press, 1997),[4] which was revised into the *NISO Z39.7-2002*.[5] ARL revised the definitions in 2003-04, incorporating the clarifications formerly provided through the *ARL Statistics* Q&A.[6] The interpretations are established through discussions within the ARL Statistics and Assessment Committee[7] and with the ARL Survey Coordinators[8] who have the responsibility for filling in the surveys. For example, see a discussion document regarding counting electronic serials.[9]

This year the existing definitions of the *ARL Statistics* data items were modified to include electronic resources, electronic journals, and e-books in the existing categories. For example, in 1999-2000 e-books were reported together with volumes held for the first time, as long as the library owned those e-books and they conformed to a print-equivalent model (see *ARL Statistics* Q&A for further clarification). In general terms, there is a sense that the *ARL Statistics*, which primarily focus on input and output measures, have served libraries adequately but in a limited way by describing the range of resources and service activities in a quantifiable manner. After a decade of testing and data collection in the *ARL Supplementary Statistics* survey, some new data elements migrated to the *ARL Statistics* in 2003-04. A new supplementary statistics survey has been developed and the data elements there focus primarily on electronic resources (e-metrics), which tend to be in flux. Development efforts tend to focus on the ARL New Measures Initiative projects[10] which emphasize service quality, impact and outcome indicators (where there is general consensus that libraries lack agreed-upon frameworks), and tools they can use to measure these aspects of their operations successfully.

SERVICE ACTIVITY TRENDS

The success of an academic library is dependent not only on the information resources it owns or licenses, but also on the services it provides. ARL collects data about public service activities such as circulation (initial and total), reference transactions, library instruction (group presentations and participants in these presentations), and interlibrary borrowing and lending (see Table 1). These data, rather than being comprehensive for the range of user-initiated library activities, represent select service areas.

Readers should take care when using service indicators for comparing institutions, because local policies can influence the level of service activities. Loan periods, for example, are usually determined by local policies; thus, a library with a shorter loan period will report a larger number of circulation transactions than will a library with a longer loan period, other things being equal. Despite the standardization efforts at the definitional level, there is wide variation at the local level in terms of the processes used.

With this in mind, it is useful to look at the trends of these select services assuming that changes due to policies and other conditions affecting measurement of services are random across institutions. For example, the median value of the ratio of total to initial circulation has been gradually increasing over the last few years, indicating that users are more likely to renew library materials, or possibly that they want the

[4] See also http://www.techstreet.com/cgi-bin/pdf/free/152592/z39-7.pdf .
[5] NISO Z39.7 – 2002 Draft Standard for Trial Use: http://www.niso.org/emetrics/.
[6] ARL Statistics Q&A: http://www.arl.org/stats/arlstat/arlstatqa.html.
[7] ARL Statistics and Measurement Committee homepage: http://www.arl.org/stats/program/meeting.html.
[8] ARL Survey Coordinators homepage: http://www.arl.org/stats/coordinator.html.
[9] Julia Blixrud, "Counting Electronic Serials: A Discussion Document," http://www.arl.org/stats/counting.html.
[10] ARL New Measures Initiative: http://www.arl.org/stats/newmeas/newmeas.html.

materials in their hands for longer periods of time. Keeping the material in the hands of the users also alleviates the space demands made on libraries by the continuing growth of their collections.

Graph 1 indicates that, since 1991, certain service areas are increasing whereas others are decreasing their activity levels. Overall, library staffing has remained roughly constant. Starting in 1996 circulation service transactions began to decline, in 1998 reference transactions began to fall, and in 2000 both categories dropped below 1991 levels for the first time. The median of group presentations increased to a new record high of 806 in 2003, and nearly matched that high with 803 in 2004-05. Meanwhile, the number of participants in those presentations continues to increase, with 13,034 participants for the typical research library. Interlibrary borrowing has also grown constantly since 1991, by an average of 1,096 transactions per year.

A variety of explanations have been voiced regarding the decline of the number of reference transactions. Many libraries are making a concerted effort to examine the changing user needs that impact reference services in general. Heavy users of library materials and services may make less use of in-person reference services than did such users in the era before the availability of online catalogs, remote access to indexing and abstracting databases, and electronic full-text resources delivered at the desktop. Often, those people who do approach reference librarians require more assistance than before. At the same time, virtual reference services are adding another dimension to the growing complexity of responding to reference questions. Libraries have instituted initiatives with a deliberate emphasis on direct contact between subject specialists and departments (shifting research consultation activity away from desk-based service). Thus, a simple count where each reference question gets a single "tally" cannot capture the varying dimensions and growing complexities of reference services. While patterns of behavior are changing and there is a decline in reference transactions, 66,300 questions per year are still made in the median ARL library.

Demand for library user education has been high recently. As seen in Graph 1, group presentations have risen by 58% since 1991, participants in these presentations by 93%, and interlibrary borrowing by 147%. Perhaps of most interest is the fact that by 2005, about one-third of the presentations conducted in a typical ARL library had been added since 1991. The typical ARL library offered 803 "teaching" sessions in 2004-05. If we assume that each session was at least an hour long, then the median ARL library offered the equivalent of 22 three-hour credit courses last year. Since a median number of 13,782 people received formal education through library instruction in a typical ARL library, those 803 "teaching" sessions averaged about 17 attendees. Information literacy has become an important program area for libraries and the Association of College and Research Libraries (ACRL) has developed widely used "Information Literacy Competency Standards for Higher Education."[11] Efforts are underway through the ARL New Measures Initiative to define how libraries contribute to student learning outcomes from a user-centered perspective.

---

[11] <http://www.ala.org/acrl/ilcomstan.html>.

Table 1
Service Trends in ARL Libraries, 1991-2005
Median Values for Time-Series Trends[12]

Year	ILL: Borrowed	Group Pres.	Participants In Group Pres.	Reference Trans.	Initial Circ.	Total Circ.	Ratio of Initial to Total Circ.	Total Staff	Total Students
(Libraries)	(103)	(84)	(82)	(80)	(36)	(80)	(34)	(105)	(103)
1991	10,397	508	7,137	128,272	296,964	509,673	1.26	271	18,290
1992	11,362	526	7,154	132,562	342,989	554,579	1.27	265	18,273
1993	12,489	616	7,688	137,580	343,293	568,628	1.32	262	18,450
1994	14,007	568	7,831	150,144	369,996	572,749	1.31	264	18,305
1995	14,472	687	8,461	148,175	347,144	578,989	1.32	267	18,209
1996	15,278	719	8,410	156,306	336,481	560,244	1.39	264	18,320
1997	16,264	687	9,218	152,164	348,157	542,438	1.37	273	18,166
1998	17,656	698	9,462	134,944	354,924	514,574	1.37	273	18,335
1999	18,942	711	9,406	129,089	300,923	514,087	1.38	277	18,609
2000	20,475	722	9,596	117,027	273,231	482,542	1.42	267	18,908
2001	21,902	669	10,121	104,744	265,195	467,277	1.48	269	19,102
2002	21,339	776	11,350	96,829	251,146	462,223	1.51	279	19,925
2003	22,146	806	12,516	91,093	248,689	479,733	1.57	277	21,132
2004	25,737	757	12,864	85,478	261,526	496,369	1.60	273	21,562
2005	25,729	803	13,782	66,300	250,971	473,216	1.58	267	22,047
Avg annual % change	6.7%	3.3%	4.8%	-4.6%	-1.2%	-0.5%	1.6%	-0.1%	1.3%

[12] Series for Interlibrary Borrowing, Group Presentations, Participants in Group Presentations, Reference Transactions, Total Circulation, and Total Students revised due to unavailable data.

Graph 1

Service Trends in ARL Libraries, 1991-2005

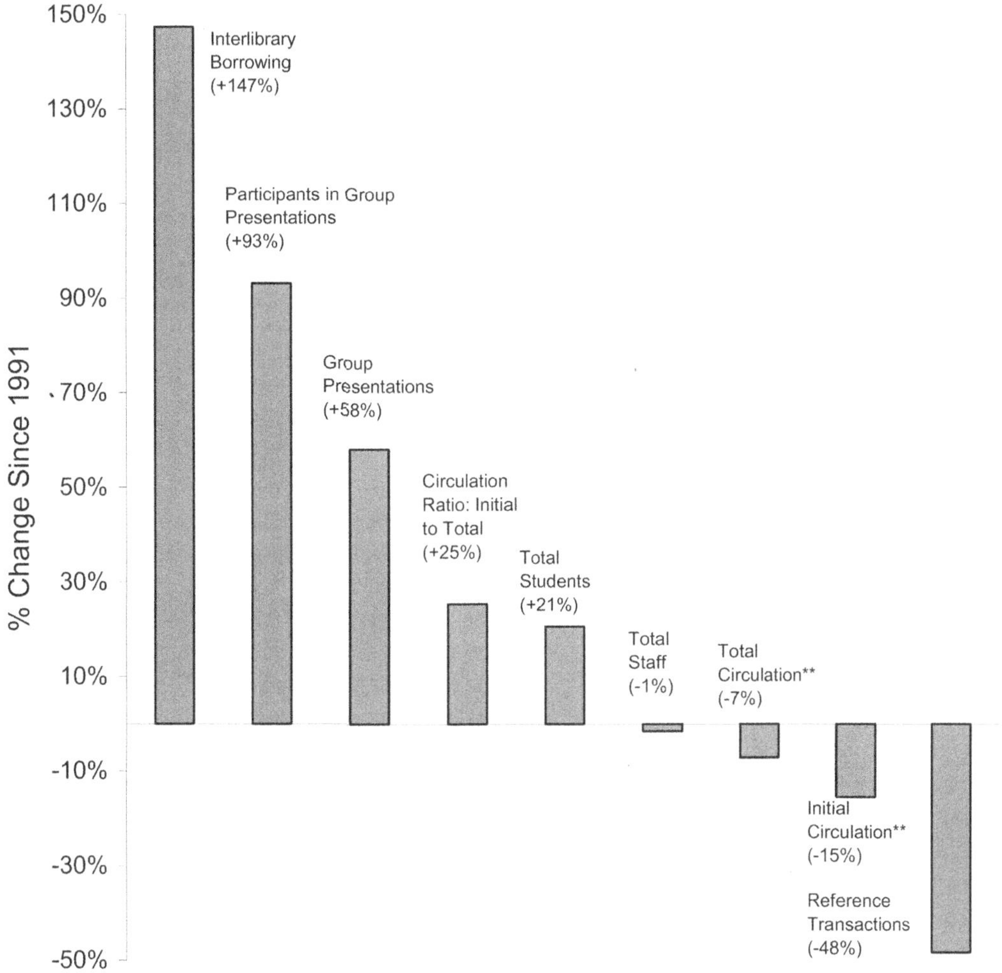

** Total Circulation includes Initial and Renewals but excludes Reserve Circulation

Source: *ARL Statistics 2004-05,* Association of Research Libraries, 2006

CHANGE IN SERIAL UNIT COSTS

The story of struggling library budgets during the 1990s had been told in terms of the "serials crisis." Serial unit costs have been increasing much faster than inflation for almost two decades, as has been shown in Table 2 and Graph 2. The electronic environment may indeed be disrupting a dysfunctional system, but it is important to keep in mind that serial subscriptions exhibit extreme inelasticity of demand (i.e., demand is very high for continuing a subscription), sometimes to the detriment of other budget lines.

Table 2
Monograph and Serial Costs in ARL Libraries, 1986-2005
Median Values for Time-Series Trends[13]

Year	Serial Unit Cost	Serial Expenditures	Monograph Unit Cost	Monograph Expenditures	Serials Purchased	Monographs Purchased
(Libraries)	(36)	(101)	(59)	(98)	(36)	(59)
1986	$89.81	$1,475,825	$29.28	$1,118,931	15,775	32,425
1987	$108.12	$1,769,353	$31.81	$1,060,754	16,514	26,204
1988	$117.41	$1,942,350	$36.06	$1,109,845	15,948	24,947
1989	$129.95	$2,097,789	$38.44	$1,093,858	15,983	26,997
1990	$135.61	$2,289,075	$40.74	$1,329,950	16,128	27,545
1991	$153.46	$2,519,065	$42.35	$1,396,566	15,962	27,659
1992	$173.69	$2,610,837	$43.99	$1,348,786	15,673	26,735
1993	$188.79	$2,917,381	$43.74	$1,284,116	15,441	24,933
1994	$203.87	$2,892,898	$44.72	$1,282,569	15,099	25,321
1995	$217.38	$3,128,181	$44.98	$1,365,046	14,320	25,695
1996	$223.98	$3,384,928	$46.73	$1,437,028	14,723	25,560
1997	$250.74	$3,610,714	$46.42	$1,457,789	14,820	28,494
1998	$252.28	$3,814,162	$47.59	$1,486,436	14,063	24,133
1999	$271.51	$4,093,793	$47.74	$1,496,687	14,192	24,398
2000	$310.62	$4,430,030	$47.59	$1,645,248	14,541	27,694
2001	$279.07	$4,610,327	$48.31	$1,848,622	13,682	30,459
2002	$289.84	$4,915,339	$50.35	$1,806,964	17,594	31,406
2003	$282.20	$5,372,822	$52.80	$1,858,280	18,115	33,177
2004	$256.01	$5,552,216	$51.24	$1,824,296	22,311	29,787
2005	$239.58	$5,933,378	$52.96	$1,776,416	22,404	30,217
Avg annual % change	5.3%	7.6%	3.2%	2.5%	1.9%	-0.4%

From the user perspective, ownership and access are interrelated; distinctions between the two may only exist inside the research library, where ownership of materials may be more closely linked to preservation functions. Data collected through LibQUAL+™ show that the demand relates to very strong user perceptions that libraries are not adequately meeting users' need of access to full runs of journal titles and delivering full-text on the desktop.[14] It is clear that some of the major scientific and technical publishers

[13] Series for all items except Monograph Expenditures were revised due to unavailable data.

[14] Bruce Thompson, Colleen Cook, and R.L. Thompson, "Reliability and Structure of LibQUAL+ Scores," *portal: Libraries and the Academy* 2 (2002): 3-12; Colleen Cook, Fred Heath, and Bruce Thompson, "Score Norms for Improving Library Service Quality: A LibQUAL+ Study," *portal: Libraries and the Academy* 2 (2002): 13-26; Fred Heath, Colleen Cook, Martha Kyrillidou, and Bruce Thompson, "ARL Index and

Graph 2

Monograph and Serial Costs in ARL Libraries, 1986-2005*

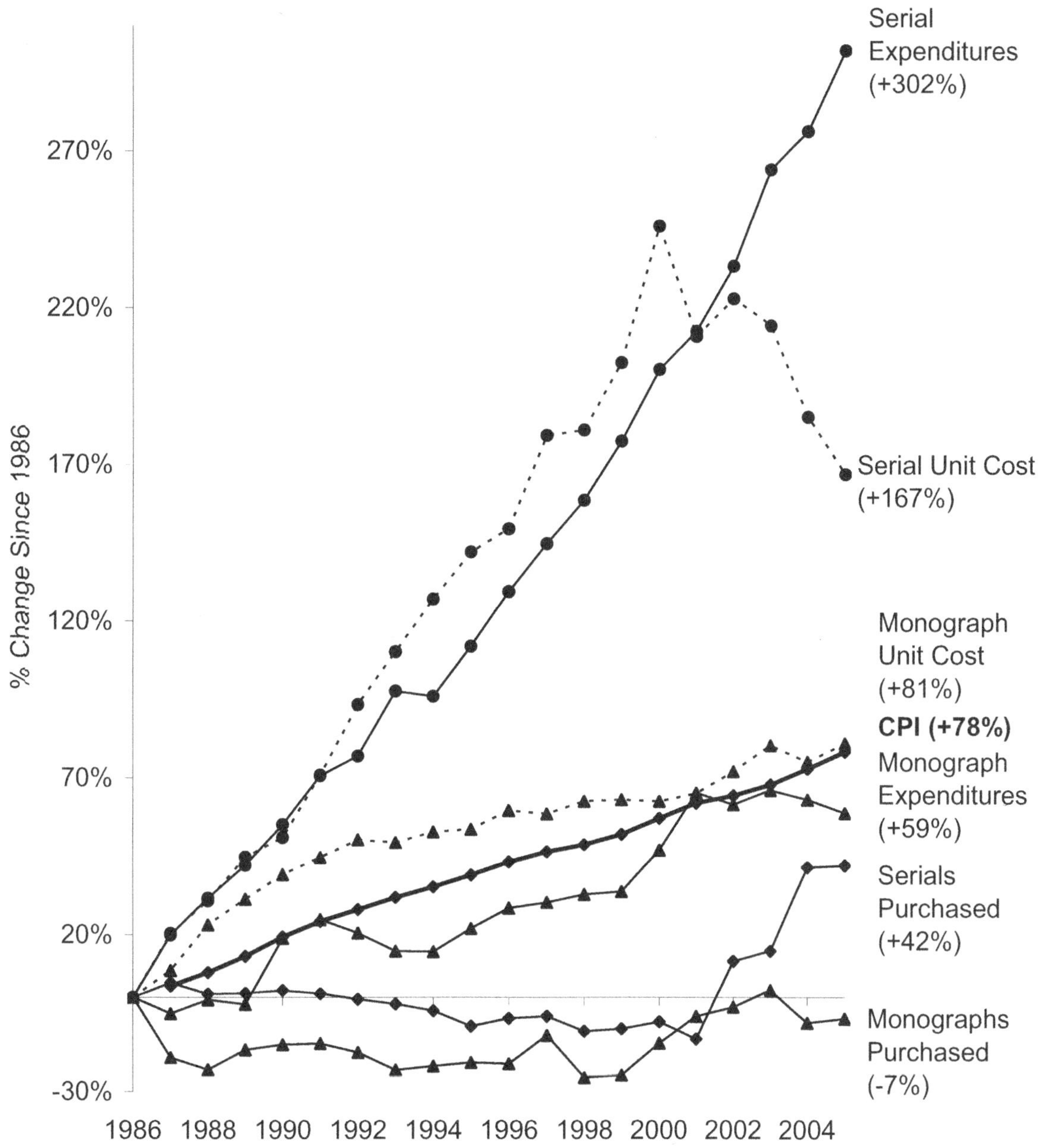

Source: ***ARL Statistics 2004-05*****, Association of Research Libraries, Washington, D.C.**
***Includes electronic resources from 1999-2000 onward.**

Other Validity Correlates of LibQUAL+ Scores," *portal: Libraries and the Academy* 2 (2002): 27-42.

have recognized this demand: for the third year in a row, the unit cost of serials declined and more and more serial subscriptions became available to libraries. A common sense explanation is that this is happening because of the proliferations of electronic journal subscriptions. It is not clear, though, whether the issues of preservation and quality control (such as assuring integrity and authenticity) of the electronic medium have been adequately addressed.

In informal conversations held with some ARL libraries, it seems that the inclusion of electronic serials in the counts of serial subscriptions purchased caused an increase in purchased serials—often, for a relatively small addition to the base subscription price, some publishers provide access to electronic resources for an additional 10 or 20% surcharge. Furthermore, the elimination of the print subscription may have resulted in discounted subscription fees for the electronic-only title; a library may have access to the electronic-only version of a journal for 80% (or some other fraction) of the print subscription price.

Other factors that may be contributing to the lower serial unit cost are consortial licensing arrangements for electronic journals, where the cost of the license is spread among participating libraries and market pressures to control the cost of serials. The Scholarly Publishing and Academic Resources Coalition (SPARC) was launched by ARL in June 1998 to promote competition in the scholarly publishing marketplace by creating "partnerships" with publishers who are developing high-quality, economical alternatives to existing high-priced publications. SPARC's activities are featured in the popular press, help increase awareness of the challenges facing libraries, and initiate specific actions aimed towards increasing competition in the largely monopolistic field of scientific scholarly publishing.[15]

During the past five years, libraries have expanded the amount of material to which they provide access by purchasing the same content in new formats and acquiring new content, often through bundling arrangements, as well as by managing the growing amount of content available through open-access mechanisms. The purchase of new and dual-format content via bundling or "big deal" arrangements[16] is probably partly responsible for the recent decline in the growth rate for serial unit costs—libraries have added serial titles to their collections at lower incremental prices. These additional titles are often duplicate subscriptions or titles the library would not otherwise purchase. Depending on the publisher's financial model, some of the additional content may be purchased or some may come bundled or "free" with a subscription to other products.[17]

---

[15] For more information, go to the SPARC homepage: http://www.arl.org/sparc/.

[16] Kenneth Frazier, "The Librarians' Dilemma: Contemplating the Costs of the 'Big Deal,'" *D-Lib Magazine* 7, no. 3 (March 2001), http://www.dlib.org/dlib/march01/frazier/03frazier.html.

[17] In the *ARL Statistics*, nonpurchased serials are not included in the calculation of serial unit cost.

Ownership and Access

Three years ago, the number of serials purchased increased above 1986 levels for the first time since 1992, and in 2004-05 the median of 22,595 subscriptions reached a new high in the history of the time series. This is most likely due to the same reasons cited above, including increased availability of electronic subscriptions and consortial arrangements. Research libraries purchased slightly more monographs in 2005 than in 2004; monographs purchased has risen above 1986 levels only once in the last nineteen years. Since 1986, the average annual increase for the serial unit cost has been 5.3%, and for the monograph unit cost 3.2%. Both are higher than the general inflation trend during the same period, and include both print and electronic resources (frequently with some duplication between the two media forms). Thus, the intellectual capital purchased by libraries is declining not only due to reduced purchasing power but also due to the degree of content duplication between the electronic and print media.

In 1986, the typical ARL library subscribed to 15,775 serials and bought 32,425 monographs for 16,660 students and 1,124 faculty. In 2005 it bought 22,404 serials and 30,217 monographs for 21,856 students and 1,355 faculty. Additionally, libraries are increasingly providing improved access without purchase of materials; Table 3 and Graph 3 show that the number of non-purchased serials received by the average ARL library increased by an annual average of 6.6% since 1986. This category consists of a number of types of serials, including government documents, electronic serials made available free of charge with the purchase of print counterparts, and open access journals. To some extent the increase is due to previously uncataloged government documents added to the library catalog as they were made available through programs such as MARCIVE. However, it is likely that emerging trends such as the open access movement[18] and institutional digital repositories[19] will continue to raise the number of serials received (not purchased) in the future. As alternative publishing models are becoming more widespread, libraries will increase the proportion of the holdings they provide access to,[20] whether they manifest themselves as serials or other new emerging forms.

Interlibrary borrowing and lending, which showed a marked growth in the last decade, seem to be stabilizing. Between 1986 and 2005, the number of items borrowed has more than tripled. ARL data show research libraries are lending 126% more items today than they did 19 years ago. In 1995-96, the cost of a lending transaction for research libraries ranged from a low of $4.87 to a high of $16.34, with an average cost of $9.48 or, with an adjustment for inflation, a 2004-05 average cost of $11.81.[21]

---

[18] Mary Case and Judith Matz, "Framing the Issue: Open Access" *ARL: A Bimonthly Report on Research Library Issues and Actions from ARL, CNI, and SPARC,* no. 226 (February 2003): 8-10, http://www.arl.org/scomm/open_access/framing.html.

[19] Clifford A. Lynch, "Institutional Repositories: Essential Infrastructure for Scholarship in the Digital Age" *ARL: A Bimonthly Report on Research Library Issues and Actions from ARL, CNI, and SPARC,* no. 226 (February 2003): 1-7, http://www.arl.org/newsltr/226/ir.html; Raym Crow, "The Case for Institutional Repositories: A SPARC Position Paper" (Washington, DC: Scholarly Publishing & Academic Resources Coalition, 2002).

[20] William Y. Arms, "Quality Control in Scholarly Publishing on the Web," *The Journal of Electronic Publishing* 8, no. 1 (August 2002); and Rob Kling, Lisa Spector, and Geoff McKim, "The Guild Model," *The Journal of Electronic Publishing* 8, no. 1 (August 2002).

[21] Mary E. Jackson, "Measuring the Performance," 2.

Table 3
Supply and Demand in ARL Libraries, 1986-2005
Median Values for Time-Series Trends[22]

Year	ILL: Borrowed	ILL Lended	Graduate Students	Teaching Faculty	Total Students	Serials Purchased	Serials Received	Monographs Purchased
(Libraries)	(103)	(103)	(104)	(101)	(103)	(36)	(36)	(59)
1986	7,047	16,092	2,327	1,124	16,684	15,775	3,318	32,425
1987	7,387	16,318	3,078	1,195	17,029	16,514	3,477	26,204
1988	7,881	17,476	3,251	1,222	17,485	15,948	3,367	24,947
1989	8,547	19,638	3,312	1,285	17,866	15,983	3,345	26,997
1990	9,595	20,837	3,314	1,278	17,745	16,128	4,304	27,545
1991	10,397	23,285	3,310	1,295	18,290	15,962	4,500	27,659
1992	11,362	22,514	3,539	1,356	18,273	15,673	5,100	26,735
1993	12,489	22,740	3,745	1,281	18,450	15,441	5,082	24,933
1994	14,007	24,039	3,794	1,289	18,305	15,099	5,518	25,321
1995	14,472	24,864	3,914	1,308	18,209	14,320	6,107	25,695
1996	15,278	25,720	3,904	1,251	18,320	14,723	5,983	25,560
1997	16,264	25,463	3,942	1,263	18,166	14,820	5,757	28,494
1998	17,656	27,223	3,880	1,247	18,335	14,063	7,111	24,133
1999	18,942	26,837	3,933	1,255	18,609	14,192	6,546	24,398
2000	20,475	27,044	3,844	1,239	18,908	14,541	7,944	27,694
2001	21,902	28,950	4,159	1,279	19,102	13,682	7,915	30,459
2002	21,339	29,021	4,067	1,251	19,925	17,594	8,769	31,406
2003	22,146	33,421	4,167	1,268	21,132	18,115	8,871	33,177
2004	25,737	33,934	4,461	1,369	21,562	22,311	9,991	29,787
2005	25,729	36,325	4,595	1,355	22,047	22,404	11,203	30,217
Avg annual % change	7.1%	4.4%	3.6%	1.0%	1.5%	1.9%	6.6%	-0.4%

However, research libraries have responsibilities for future generations; cost considerations of short-term use are not adequate to ensure research level collections, whether in digital or analog formats. According to a report on collections and access issued by ARL, "developments in digital technology, the introduction of the Web and the Internet, and new methods of creating, sharing, and using knowledge have changed dramatically the traditionally understood definitions of library collections and access services. Building collections and creating access to them are no longer achieved just within the walls of the library. Broadly defined, collections and access responsibilities are no longer distinct spheres within research libraries. Collections and access responsibilities are inextricably linked—with each other, with other functions in the parent institutions, and, indeed, with other institutions. This interdependent and fluid environment presents challenges but, more importantly, it presents opportunities for librarians to take leadership roles in creating new information services in support of research and learning and thereby diffuse the library throughout the institution."[23]

---

[22] All time series in this table were revised due to unavailable data.

[23] ARL Collections & Access Issues Task Force, "Collections & Access for the 21st-Century Scholar: Changing Roles of Research Libraries," *ARL: A Bimonthly Report on Research Library Issues and Actions from ARL, CNI, and SPARC,* no. 225 (December 2002), http://www.arl.org/newsltr/225/.

Graph 3

Supply and Demand in ARL Libraries, 1986-2005*

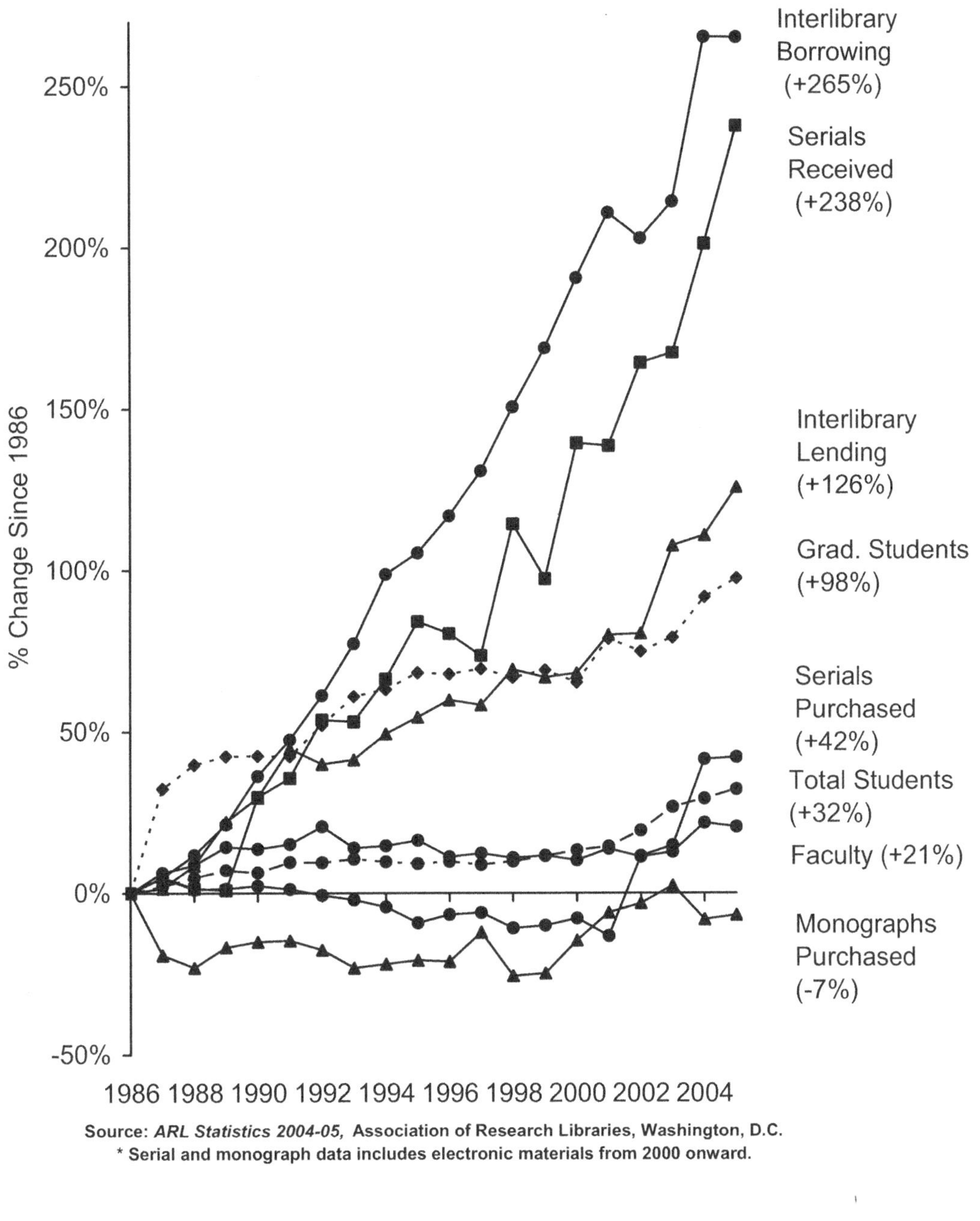

Source: ***ARL Statistics 2004-05,*** **Association of Research Libraries, Washington, D.C.**
*** Serial and monograph data includes electronic materials from 2000 onward.**

EXPENDITURE TRENDS

Library material budgets have risen sharply in order to sustain serial expenditures. Operating expenditures, including many automation expenditures, have doubled since 1986, but during the last five years have remained at relatively stable levels of investment—slightly more than $2 million for the typical ARL library. The total salary expenditures median has increased only slightly over the past four years, indicative of the difficulties libraries are having in recruitment and the mechanisms they are trying to develop to compensate for the historically low salaries paid to their employees. Monograph expenditures continue to increase, but at a much slower pace to accommodate the sharply increasing serial expenditures. The annual consumer price index (CPI), included in Graph 4, provides a comparative reference for the increases in library expenditures.

Table 4
Expenditure Trends in ARL Libraries, 1986-2005
Median Values for Time-Series Trends – Unadjusted dollar figures[24]

Year	Library Materials	Serial Exp.	Monograph Exp.	Total Salary	Operating Exp.	Total	CPI
(Libraries)	(105)	(101)	(98)	(104)	(104)	(105)	
1986	$2,705,297	$1,475,825	$1,118,931	$4,011,436	$1,111,914	$8,361,092	109.6
1987	$3,058,479	$1,769,353	$1,060,754	$4,361,646	$1,180,167	$8,990,001	113.6
1988	$3,369,896	$1,942,350	$1,109,845	$4,618,335	$1,198,674	$9,557,623	118.3
1989	$3,577,405	$2,097,789	$1,093,858	$5,236,292	$1,364,558	$10,183,315	124.0
1990	$3,903,358	$2,289,075	$1,329,950	$5,469,333	$1,386,618	$11,241,022	130.7
1991	$4,064,344	$2,519,065	$1,396,566	$5,885,814	$1,445,735	$11,990,794	136.2
1992	$4,156,510	$2,610,837	$1,348,786	$6,050,222	$1,390,245	$12,249,150	140.3
1993	$4,316,674	$2,917,381	$1,284,116	$5,962,470	$1,561,122	$12,265,696	144.5
1994	$4,572,276	$2,892,898	$1,282,569	$6,047,803	$1,676,701	$12,767,348	148.2
1995	$4,715,203	$3,128,181	$1,365,046	$6,312,770	$1,853,586	$13,171,893	152.4
1996	$5,126,482	$3,384,928	$1,437,028	$6,664,021	$1,997,233	$13,870,378	156.9
1997	$5,562,742	$3,610,714	$1,457,789	$6,893,582	$2,039,957	$14,526,674	160.5
1998	$5,795,223	$3,814,162	$1,486,436	$7,163,979	$2,072,903	$15,329,371	163.0
1999	$6,232,365	$4,093,793	$1,496,687	$7,476,532	$2,069,887	$16,737,261	166.6
2000	$6,744,281	$4,430,030	$1,645,248	$7,811,403	$1,991,852	$17,221,441	172.2
2001	$7,322,507	$4,610,327	$1,848,622	$8,106,666	$2,280,493	$17,620,048	177.5
2002	$7,599,249	$4,915,339	$1,806,964	$8,488,255	$2,136,616	$18,456,038	180.1
2003	$8,273,171	$5,372,822	$1,858,280	$8,813,191	$2,073,913	$19,030,188	183.9
2004	$8,286,431	$5,552,216	$1,824,296	$9,015,741	$2,274,878	$19,953,776	189.4
2005	$8,801,962	$5,933,378	$1,776,416	$9,268,364	$2,243,592	$20,663,012	195.4
Avg annual % change	6.4%	7.6%	2.5%	4.5%	3.8%	4.9%	3.1%

[24] Time series for Serial Expenditures, Total Salary Expenditures, and Operating Expenditures were revised due to unavailable data.

Graph 4

Expenditure Trends in ARL Libraries, 1986-2005

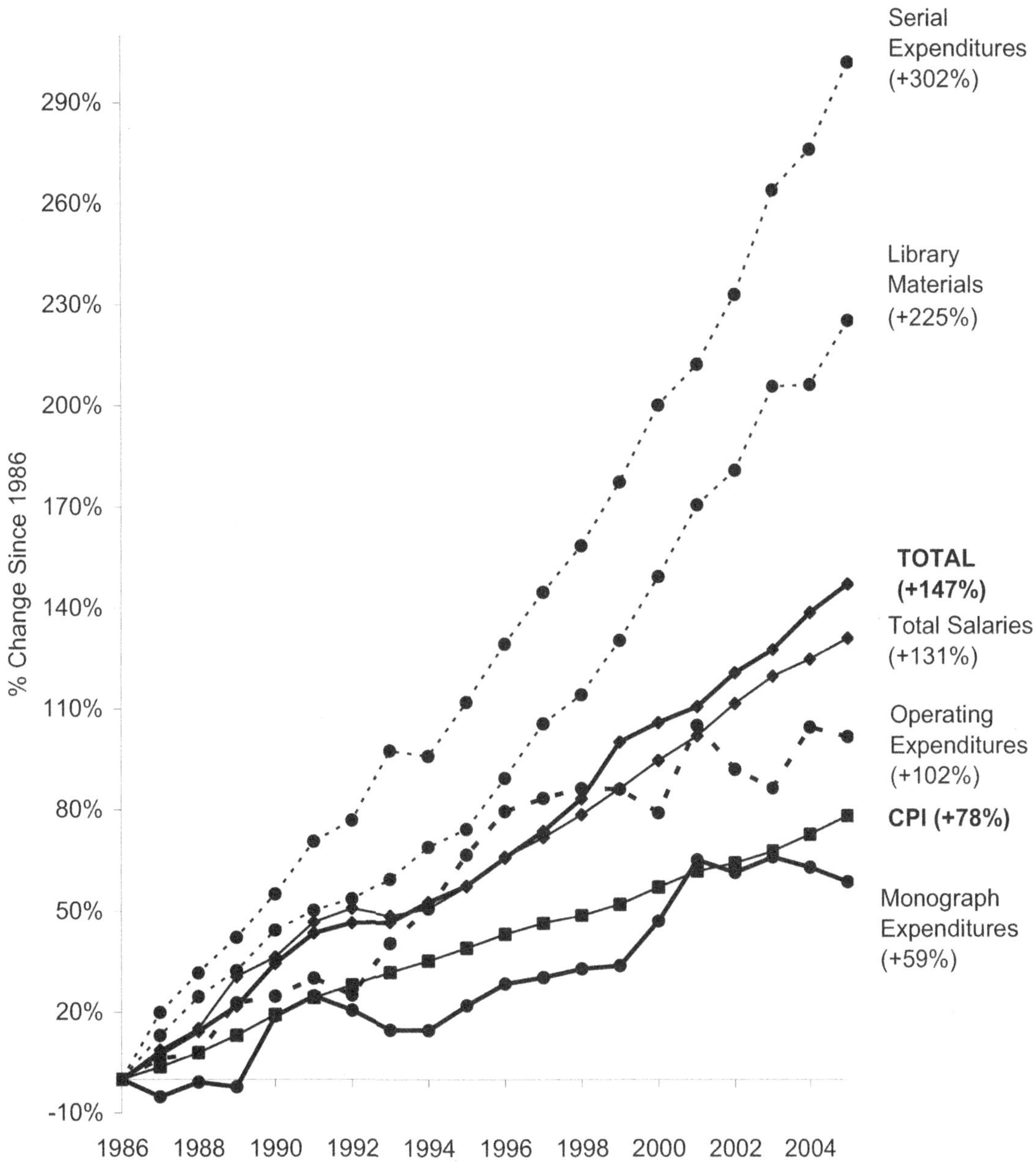

Source: *ARL Statistics 2004-05,* Association of Research Libraries, Washington, D.C.

Table 5, "Resources per Student in ARL University Libraries," shows that per-student borrowing activity through interlibrary loan has continued to increase at an annual average rate of 5.8% since 1986. Further, libraries are borrowing three times more items on a per-student basis than they did 19 years ago.[25] Table 5 also shows per-student acquisition levels for both serials and monographs; while monographs purchased per student dropped in 2004-05, the number of serials purchased per student increased by 12%. As a result, ARL libraries acquired 37% fewer monographs per student in 2005 than in 1986, but 30% more serials. ARL libraries acquired 1,100 serial subscriptions and 1,360 monographs per 1,000 students and the median number of volumes added was 3.27 per student, compared to the 4.14 volumes added per student in 1986. Libraries also reported fewer staff per student in 2005 as compared to 1986: in 2005, there was a median number of 11.9 total staff per 1,000 students, compared to the 1986 figure of 16 per 1,000.

Table 5
Resources per Student in ARL University Libraries, 1986-2005
Median of Ratio Values for Time-Series Trends[26]

Year	ILL: Borrowed	Volumes Held	Volumes Added (Gross)	Total Staff	Serials Purchased	Monographs Purchased
(Libraries)	(103)	(104)	(104)	(104)	(36)	(59)
1986	0.42	123.27	4.14	0.0159	0.85	2.15
1987	0.43	125.78	3.96	0.0165	0.84	1.86
1988	0.44	128.65	3.82	0.0161	0.85	1.65
1989	0.50	130.32	3.79	0.0163	0.86	1.75
1990	0.55	133.78	3.85	0.0156	0.86	1.83
1991	0.62	137.41	3.93	0.0154	0.85	1.79
1992	0.62	137.68	3.94	0.0153	0.84	1.54
1993	0.67	146.08	3.74	0.0152	0.80	1.46
1994	0.74	151.43	4.06	0.0153	0.79	1.54
1995	0.80	154.27	4.40	0.0147	0.83	1.56
1996	0.87	158.62	3.74	0.0146	0.80	1.44
1997	0.88	159.87	4.01	0.0142	0.77	1.52
1998	0.96	159.52	4.06	0.0145	0.73	1.42
1999	1.04	160.34	4.00	0.0142	0.77	1.42
2000	1.13	158.32	3.79	0.0141	0.77	1.63
2001	1.17	157.55	3.79	0.0136	0.79	1.65
2002	1.15	156.93	3.85	0.0131	0.82	1.54
2003	1.14	156.95	3.49	0.0126	0.85	1.46
2004	1.20	154.75	3.42	0.0121	0.98	1.53
2005	1.23	158.79	3.27	0.0119	1.10	1.36
Avg annual % change	5.8%	1.3%	-1.2%	-1.5%	1.4%	-2.4%

[25] This overall trend should not be interpreted rigidly, as it negates the varying experiences of individual libraries.

[26] All time-series in this table have been revised due to unavailable data.

Graph 5

Resources per Student in ARL University Libraries, 1986-2005

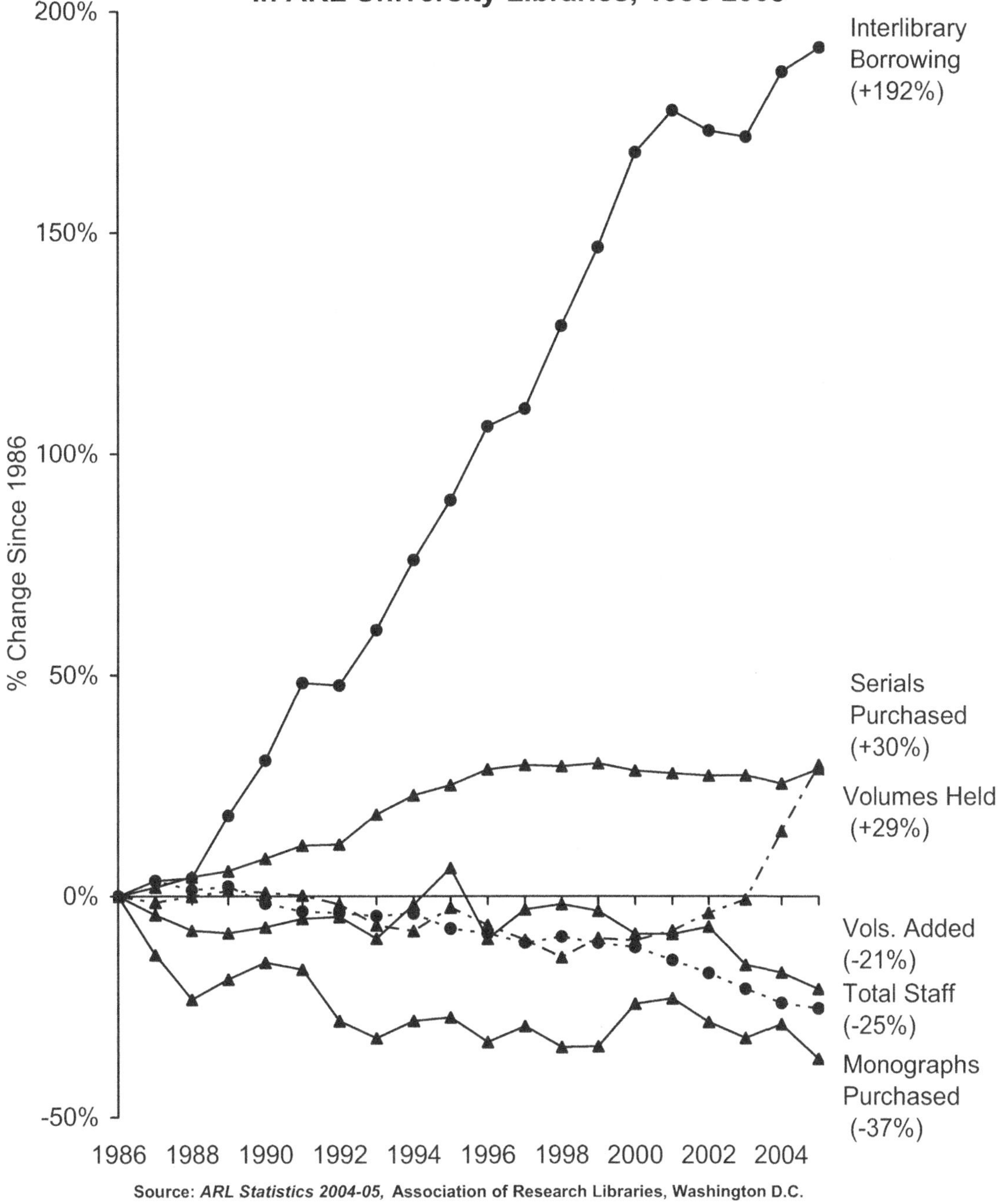

Source: *ARL Statistics 2004-05,* Association of Research Libraries, Washington D.C.

EXPENDITURES FOR ELECTRONIC MATERIALS

Starting in 2003-04, the *ARL Statistics* collected data on several items which previously had been collected only in the *ARL Supplementary Statistics,* all of which are listed in Table 6. Expenditures for electronic resources have all increased substantially since their introduction into the *Supplementary Statistics,* but none more than expenditures for electronic serials. E-serials expenditures were just $11,847,577 from 63 reporting libraries in 1994-95; in 2004-05, 108 libraries reported almost $330 million spent. These data are especially useful because they reflect monies spent on all electronic serials, while the *ARL Statistics* categories of "serials purchased" and "serials expenditures" include only those journals which provide full-text electronic versions to their subscribers. The Expenditures for Electronic Serials time series may be viewed as an alternative to the Serials Purchased series, both figures reflecting in their own way the influence the electronic serial is gaining in the modern research library.

Table 6
Electronic Materials Expenditures
In ARL University Libraries, 2004-05

	Sum	Number Reporting
Expenditures for Computer Files (one-time/monographic purchase)	38,744,076	104
Expenditures for Electronic Serials	328,166,027	108
Expenditures for Bibliographic Utilities, Networks, etc. (Library)	25,203,164	105
Expenditures for Bibliographic Utilities, Networks, etc. (External)	16,082,790	81
Expenditures for Hardware and Software	91,790,199	106
Expenditures for Document Delivery/Interlibrary Loan	12,951,510	106
Staffed Library Service Points	2,732	110
Library Service Hours	117.10 [27]	110

Furthermore, not only have electronic materials expenditures grown sharply in the past decade, but they have grown at a rate far exceeding that of library materials expenditures overall. As shown in Graph 6 (page 22), in every year of the last decade electronic materials expenditures have grown sharply, anywhere between three and ten times faster than materials expenditures have. The average ARL university library now spends more than 37% of its materials budget on electronic materials (Table 7), and fifteen ARL libraries report that they spent more than 50% of their materials budget on electronic materials (see Rank Order Table 20).

[27] Figure is not a sum, but instead it reflects average service hours per reporting institution.

TABLE 7
ELECTRONIC RESOURCES AND MATERIALS EXPENDITURES IN ARL UNIVERSITY LIBRARIES, 1992-2005

	1992-93	1993-94	1994-95	1995-96	1996-97	1997-98	1998-99	1999-2000*	2000-01	2001-02*	2002-03	2003-04	2004-05
a. Computer File Expenditures (monographic/onetime)													
Total	14,147,625	20,132,553	22,030,727	24,639,822	8,013,055	11,189,103	10,848,219	14,727,984	15,297,096	16,748,194	23,275,683	32,098,404	38,744,076
Average	172,532	236,854	247,536	262,126	87,098	122,957	121,890	161,846	159,345	167,482	225,978	314,690	372,539
Median	148,158	212,936	217,988	219,178	47,932	52,311	54,024	98,657	72,070	82,566	111,266	191,148	210,576
N		85	89	94	92	91	89	91	96	100	103	102	104
b. Electronic Serial Expenditures													
Total	N/A	N/A	11,847,577	15,170,971	40,956,696	49,497,141	67,124,554	84,343,868	117,415,618	154,418,679	205,300,292	269,601,241	328,166,027
Average	N/A	N/A	188,057	194,500	401,536	494,971	639,281	818,873	1,118,244	1,429,803	1,849,552	2,450,920	3,038,574
Median	N/A	N/A	156,754	172,805	355,922	426,722	571,790	736,317	992,067	1,272,965	1,649,361	2,348,463	2,824,962
N			63	78	102	100	105	103	105	108	111	110	108
c. Total Electronic Resources (Total a+b)													
Total	14,147,625	20,132,553	33,878,304	39,810,793	50,512,984	60,686,244	77,972,773	99,071,852	132,712,714	171,166,873	228,575,974	301,699,645	366,910,103
Average	172,532	236,854	349,261	394,166	485,702	594,963	742,598	943,541	1,252,007	1,556,062	2,059,243	2,718,015	3,366,147
Median	148,158	212,936	278,404	332,128	420,741	495,011	645,495	931,210	1,129,298	1,377,874	1,775,865	2,705,847	3,144,841
N	82	85	97	101	104	102	105	105	106	110	111	111	109
Total Library Materials Expenditures [28]													
Total	393,271,073	425,287,651	526,496,347	571,145,986	642,123,715	664,600,663	727,623,160	773,321,519	828,778,808	910,930,849	950,275,167	1,016,121,605	1,031,619,722
Average	4,795,989	5,003,384	5,427,797	5,654,911	6,174,266	6,515,692	6,929,744	7,364,967	7,818,668	8,281,189	8,561,038	9,154,249	9,464,401
Median	4,242,887	4,527,122	4,714,384	4,975,353	5,529,606	5,643,070	5,991,177	6,545,146	7,028,134	7,566,727	7,707,153	8,276,175	8,662,668
N	82	85	97	101	104	102	105	105	106	110	111	111	109
Electronic Resources Expenditures as a Percent of Total Materials Expenditures													
Average	3.60	4.75	6.39	6.83	7.76	8.85	10.56	12.88	16.25	19.60	25.02	31.33	37.46
Median	N/A	4.45	5.33	6.42	7.51	8.29	10.18	12.75	14.80	18.15	22.01	29.81	37.53
N	82	85	97	101	104	102	105	105	106	110	111	111	109
Expenditures for Bibliographic Utilities, Networks, etc. (External)													
Total	N/A	N/A	N/A	N/A	$3,827,348	4,695,737	7,442,962	9,523,348	14,655,078	20,373,560	21,470,716	17,420,520	16,082,790+
Average	N/A	N/A	N/A	N/A	$136,691	142,295	201,161	250,614	311,810	424,449	438,178	335,010	349,626+
Median	N/A	N/A	N/A	N/A	$120,096	128,795	145,280	204,598	198,289	336,690	250,000	94,837	149,396+
N					28	33	37	38	47	48	49	52	46+

[28] Figures reflect Materials Expenditures only from those institutions that reported nonzero figures for Total Electronic Resources.

* In a recent review of past Supplementary Statistics data, some figures previously published in these columns were found to be incorrect and subsequently revised.

+ Includes only nonzero responses, to be consistent with past Supplementary Statistics reporting. Statistics that include zeroes can be found on page 51.

Graph 6
Yearly Increases in Average:
Electronic Resources vs. Total Materials Expenditures, 1993-2005

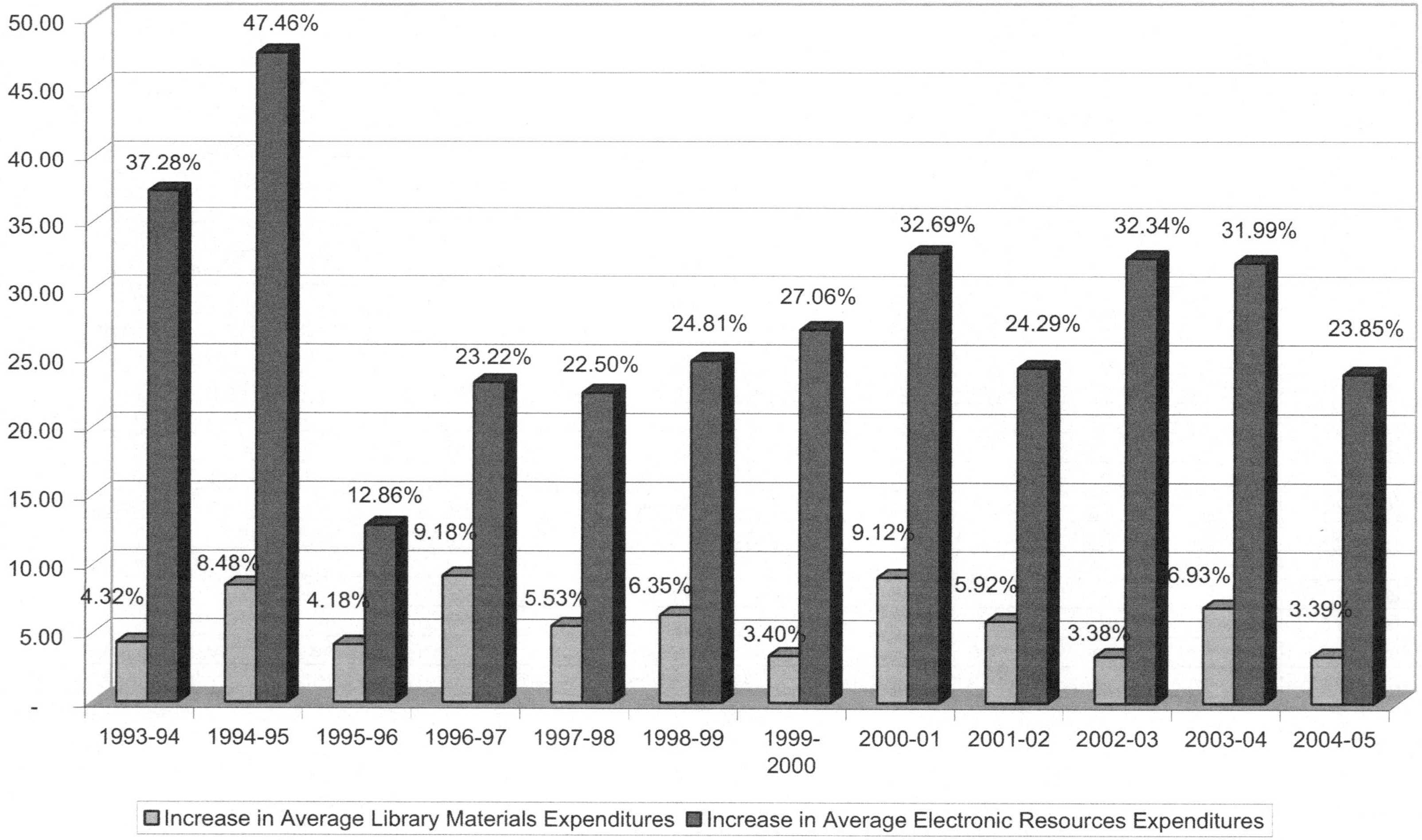

THE TRENDS CONTINUE

The Web has revolutionized the way libraries are delivering services, enabling them to offer more value ranging from remote access to online catalogs, indexing and abstracting tools, and full-text resources delivered at the user's desktop. The delivery of new and innovative services through digitization projects and distance learning technologies is transforming the brick-and-mortar library model to a virtual model.

Higher education is changing. Recently the well-known classification by the Carnegie Foundation for the Advancement of Teaching shows that institutions are increasingly described in many different ways, based on different characteristics. "The Carnegie Classifications has traditionally grouped institutions by degrees offered, so that doctoral institutions were in one group and community colleges in another, and so forth. The new classifications[29] take a very different approach. Institutions are grouped (multiple times) based on what is taught, to whom, and in what setting. The old system — with some revisions — will still be used."[30] The recent Spellings report calls for further evidence of accountability, accessibility, affordability and quality.[31]

Library roles are being redefined as the research and academic community undergoes profound changes. The ARL Statistics and Measurement program and its advisory Statistics and Assessment Committee continue to look for new ways to describe and measure the performance of research libraries and their contributions to teaching, research, learning, and community service. In a period of rapid technological change and fluctuation, the information gathered here represents only a basic and rudimentary picture of the major trends affecting research libraries, their resources, and their use. The challenge of describing libraries at a time when Google™ promises to digitize the largest research libraries of the world is a formidable one.

This data compilation does not assess the quality of a library in meeting user needs, nor does it provide a complete picture of investments in electronic resources and other innovations. Answers to these questions can only be found by other measures, such as asking library users about their real needs and then designing better service delivery systems. ARL is engaged in a variety of projects that aim to assess the library's impact on teaching, learning, and research, as well as the ability of libraries to control costs and add value to the services they provide. William Crowe captured the importance of the increased attention ARL libraries gave to measurement and assessment during the last decade by characterizing the movement as a "move beyond the rearview mirror approach."[32]

Library assessment is gaining in momentum and importance within libraries[33] with a multiplicity of methods and tools now available from ARL to libraries including LibQUAL+®, MINES for Libraries™, DigiQUAL™, in addition to the regular ARL Statistics. Resource investments in electronic products are tracked through the *ARL Supplementary Statistics* and other pilot projects. The ARL E-Metrics pilot resulted in a revised *ARL Supplementary Statistics* in 2003-04.

---

[29] http://www.carnegiefoundation.org/classifications/
[30] *Inside Higher Education,* November 21, 2005.
[31] The Secretary of Education, Margaret Spellings, has issued the Spellings Commission Report on higher education, found at http://www.ed.gov/about/bdscomm/list/hiedfuture/reports/pre-pub-report.pdf
[32] William J. Crowe, "The End of History? Reflections on a Decade" *ARL: A Bimonthly Report on Research Library Issues and Actions from ARL, CNI, and SPARC* 226 (February 2003): 12-13, <http://www.arl.org/newsltr/226/endofhistory.html>.
[33] Steve Hiller, Martha Kyrillidou and Jim Self, "Assessment in North American research libraries: a preliminary report card." *Performance Measurement and Metrics* 7 (2) (2006): 100-106.

StatsQUAL™ is a gateway to library assessment tools that describe the role, character, and impact of physical and digital libraries on research, teaching and learning including both new and innovative tools as well as the traditional descriptive data collected through ARL Statistics. StatsQUAL™ is an effort to present these tools in a single powerful interactive framework that integrates and enhances data mining and presentation both within and across institutions. Enhancements and improvements in the StatsQUAL™ interface are iterative and evolving as we move into the future.

In conclusion, those using the *ARL Statistics* to compare individual libraries should consult the extensive "Footnotes" section and the symbols in the "ARL Library Data Tables." Although definitions used in the *Statistics* aim for consistency, differing reporting practices do exist. To aid comparability, Canadian library expenditures are expressed in U.S. dollars, at the rate of 1.24971 Canadian dollars to one U.S. dollar. This exchange rate is the average monthly noon exchange rate published in the *Bank of Canada Review* for the period July 2004–June 2005. Expenditures reported in Canadian dollars are given in the "Footnotes."

The quantitative rank-order tables presented in this publication are not indicative of performance and outcomes and should not be used as measures of library quality. In comparing any individual library to ARL medians or to other libraries, one must be careful to make such comparisons within the context of differing institutional goals and local characteristics.

Martha Kyrillidou and Mark Young
Association of Research Libraries
October 31, 2006

ARL STATISTICS INTERACTIVE EDITION ON THE WORLD WIDE WEB

The *ARL Statistics* are available through an interactive edition on the Web, supplementing previously available data from 1992-93. Maintained and supported by the Geospatial and Statistical Data Center at the University of Virginia, this site offers researchers access to *ARL Statistics* with documentation.[1]

The site offers maps and graphics depicting current library trends. For the particularly adventurous, the site links to an interactive graphics and statistics page, allowing users to manipulate their data queries. One may chose univariate, bivariate, or multivariate statistics in addition to a vast array of selections with which to customize a query. The interactive pages also include options for regional aggregates, customized samples of observations, simple rank order tabulations of data, and the ability to create ratios of any two variables.

The site also includes membership index data (volumes held, volumes added gross, current serials, total library expenditures, and total staff), as published in the *Chronicle of Higher Education* each spring, which can be displayed as numbers or graphs. A link provides an explanation of how the ARL membership index is used in assessing new numbers and the table of formulas for calculating the index since 1986 is also included. These and other capabilities are located at:

<http://fisher.lib.virginia.edu/arl/>

The ARL Statistics datafiles since 1908 can be retrieved in spreadsheet format from the ARL FTP site:

<ftp://www.arl.org/stat/machine/>

The latest ARL membership index data can be found at:

<http://www.arl.org/stats/factor.html>

The ARL Statistics and Measurement Program Web site is located at:

<http://www.arl.org/stats/>

---

[1] For more information on the *ARL Statistics* Interactive Edition, see Martha Kyrillidou, "To Describe and Measure the Performance of North American Research Libraries," *IFLA Journal* 27 (2001): 257-63.

ARL Library Data Tables, 2004-05

ARL Library Data Tables, 2004-05
Collections

(Survey Question #) INSTITUTION	Notes	Volumes in Library (1)	Volumes Added (Gross) (1.b.i)	Volumes Added (Net) (1.b)	Monographs Purchased (Volumes) (2)	Current Serials Purchased (4.a)	Current Serials Received (4.b)	Current Serials Total (4)
ALABAMA	BGLM+	2,518,290	57,476	51,312	50,909	22,663	10,133	32,796
ALBERTA	GLM+	6,190,396	166,508	163,126	UA/NA	28,257	13,690	41,947
ARIZONA	BGLM+	5,340,726	103,576	68,869	49,672	UA/NA	UA/NA	33,575
ARIZONA STATE	GL+	4,156,732	72,240	65,174	60,160	21,282	11,840	33,122
AUBURN	G+	2,918,859	37,370	14,885	23,579	10,792	18,563	29,355
BOSTON	GLM+	2,427,253	47,070	30,891	17,181	UA/NA	UA/NA	33,983
BOSTON COLLEGE	BGL+	2,124,242	55,116	47,398	UA/NA	37,171	15,167	52,338
BRIGHAM YOUNG	BGL+	3,621,577	86,125	83,372	43,681	18,426	9,379	27,805
BRITISH COLUMBIA	GLM+	5,523,194	85,516	85,516	UA/NA	UA/NA	UA/NA	53,114
BROWN	GM+	3,509,710	49,800	48,292	33,405	21,888	10,744	32,632
CALIFORNIA, BERKELEY	L+	9,985,905	192,944	156,935	UA/NA	UA/NA	UA/NA	93,256
CALIFORNIA, DAVIS	GLM+	3,490,356	57,786	46,701	UA/NA	UA/NA	UA/NA	41,273
CALIFORNIA, IRVINE	BGM+	2,439,525	46,193	41,070	29,659	20,649	5,844	26,493
CALIFORNIA, LOS ANGELES	GLM+	8,064,896	110,858	76,711	65,953	UA/NA	UA/NA	78,236
CALIFORNIA, RIVERSIDE	BG+	2,368,843	63,958	63,317	35,287	19,739	8,412	28,151
CALIFORNIA, SAN DIEGO	GM+	3,149,836	109,064	78,375	43,233	20,543	10,935	31,478
CALIFORNIA, SANTA BARBARA	+	2,837,752	21,565	19,328	28,698	22,084	18,375	40,459
CASE WESTERN RESERVE	GLM+	2,471,504	26,268	18,773	13,840	12,235	14,730	26,965
CHICAGO	BGLM+	7,363,549	165,351	157,625	98,811	UA/NA	UA/NA	43,396
CINCINNATI	GLM+	3,123,318	87,762	72,270	42,385	36,311	1,357	37,668
COLORADO	BG+	3,554,826	78,213	69,844	35,690	10,656	17,654	28,310
COLORADO STATE	G+	2,026,508	67,518	59,473	30,217	19,993	8,844	28,837
COLUMBIA	LM+	9,277,042	223,923	191,183	249,680	59,936	29,516	89,452
CONNECTICUT	GLM+	3,247,976	42,220	36,545	26,317	30,295	19,771	50,066
CORNELL	GL+	7,644,371	136,758	107,226	UA/NA	65,923	14,710	80,633
DARTMOUTH	GM+	2,474,288	44,517	39,500	26,948	42,880	5,486	48,366
DELAWARE	G+	2,667,242	46,237	43,688	28,892	7,518	5,012	12,530
DUKE	BLM+	5,560,966	73,037	67,038	59,547	31,892	1,071	32,963
EMORY	LM+	3,107,525	87,371	79,222	40,009	UA/NA	UA/NA	36,898
FLORIDA	BGLM+	4,137,528	64,670	62,238	47,038	68,194	2,688	70,882
FLORIDA STATE	BLM+	2,889,810	33,874	14,822	18,726	40,475	1,601	42,076
GEORGE WASHINGTON	BGLM+	2,167,871	54,646	38,539	22,076	15,855	3,054	18,909
GEORGETOWN	BGLM	2,473,208	77,272	64,934	34,936	24,115	5,139	29,254
GEORGIA	GL+	4,179,316	77,062	64,705	46,157	29,047	19,180	48,227
GEORGIA TECH	G+	2,410,964	40,950	40,139	5,827	12,506	18,293	30,799
GUELPH	G+	1,559,078	19,453	3,693	14,261	11,317	500	11,817
HARVARD	GLM+	15,555,533	429,344	132,354	UA/NA	UA/NA	UA/NA	98,773
HAWAII	G+	3,410,468	54,437	54,437	20,707	19,789	6,816	26,605
HOUSTON	BGL+	2,357,009	44,841	39,721	31,384	18,483	3,362	21,845
HOWARD	BLM	2,388,073	32,345	31,176	3,689	9,764	1,553	11,317
ILLINOIS, CHICAGO	GM+	2,284,649	53,000	48,017	32,712	26,350	315	26,665
ILLINOIS, URBANA	GL+	10,370,777	179,898	178,882	77,420	UA/NA	UA/NA	88,650

UA/NA – Unavailable or Not Applicable
L - Includes Law library
M- Includes Medical library

+ - See Footnotes
B - Basis of volume count is bibliographic
G - Government documents not included in serials count

ARL Library Data Tables, 2004-05
Collections

Microform Units (6)	Government Documents (7)	Computer Files (8)	Archives and Manus. (9)	Carto-graphic Materials (10)	Graphic Materials (11)	Audio Materials (12)	Video and Film (13)	(Survey Question #) INSTITUTION
4,014,614	930,419	5,934	31,685	305,192	193,475	22,953	5,238	ALABAMA
3,716,844	0	9,029	22,048	1,494,165	UA/NA	UA/NA	UA/NA	ALBERTA
6,210,417	UA/NA	12,025	9,521	264,384	352,320	41,910	24,909	ARIZONA
7,602,664	797,044	7,069	23,583	228,330	1,207,046	63,770	37,068	ARIZONA STATE
2,670,252	778,424	2,541	10,120	150,123	75,327	10,247	4,513	AUBURN
4,557,897	UA/NA	1,663	44,791	1,700	17,700	53,795	1,148	BOSTON
4,050,845	214,674	446	7,285	17,779	125,577	16,465	14,140	BOSTON COLLEGE
3,825,198	414,836	3,029	38,434	252,440	1,154,991	174,244	21,170	BRIGHAM YOUNG
5,117,219	UA/NA	45,648	11,892	578,776	520,693	94,542	15,354	BRITISH COLUMBIA
1,928,655	0	5,493	9,968	143,031	949,234	73,206	6,785	BROWN
6,825,994	98,774	14,710	UA/NA	426,658	4,570,694	86,651	39,083	CALIFORNIA, BERKELEY
4,203,651	74,595	10,197	16,959	258,401	234,940	14,548	2,357	CALIFORNIA, DAVIS
2,884,821	170,730	9,801	3,781	10,964	71,656	11,442	9,424	CALIFORNIA, IRVINE
6,166,293	69,577	13,331	58,038	631,715	4,143,378	258,200	26,377	CALIFORNIA, LOS ANGELES
1,772,053	297,740	8,454	3,119	109,222	20,921	17,652	6,149	CALIFORNIA, RIVERSIDE
3,201,291	0	19,018	12,533	143,627	313,552	161,728	15,240	CALIFORNIA, SAN DIEGO
3,759,140	413,314	12,586	16,381	3,895,944	1,824	127,977	8,418	CALIFORNIA, SANTA BARBARA
2,548,156	24,675	7,537	1,826	64,011	72,556	29,759	19,620	CASE WESTERN RESERVE
3,011,073	UA/NA	8,794	31,974	427,048	0	48,137	3,546	CHICAGO
3,340,167	38,528	12,147	25,154	168,928	297,605	56,345	20,562	CINCINNATI
6,802,839	973,069	11,632	43,824	212,460	367,876	65,721	17,331	COLORADO
2,520,094	286,420	9,440	3,862	67,314	4,344	5,482	456	COLORADO STATE
6,176,325	193,534	4,583	232,205	138,281	780,832	93,165	26,574	COLUMBIA
4,397,856	7,922	3,554	25,004	232,205	2,698	56,973	10,417	CONNECTICUT
8,204,567	UA/NA	15,989	62,930	249,000	36,867	113,162	25,658	CORNELL
2,567,627	0	10,016	19,954	187,824	498,684	34,778	11,951	DARTMOUTH
3,419,588	445,614	33,781	4,595	126,245	812	1,903	16,012	DELAWARE
4,230,496	1,246,974	27,219	51,827	131,839	876,409	49,372	64,735	DUKE
5,368,365	352,310	9,820	19,440	16,620	2,908	42,239	30,161	EMORY
7,680,340	1,223,826	12,944	8,426	837,029	92,353	26,243	11,492	FLORIDA
8,057,015	852,152	9,228	4,736	173,621	1,195	58,822	17,026	FLORIDA STATE
2,864,551	0	542	19,027	18,676	169,938	10,359	16,543	GEORGE WASHINGTON
3,877,478	9,385	6,450	16,794	1,809	368,678	17,144	18,016	GEORGETOWN
6,537,902	1,055,763	16,973	54,424	611,828	1,629,559	123,919	138,650	GEORGIA
4,495,012	0	15,080	4,811	197,407	107,564	1,866	4,908	GEORGIA TECH
2,305,363	0	2,672	11,692	73,511	6,560	11,437	7,272	GUELPH
9,903,096	UA/NA	UA/NA	UA/NA	UA/NA	UA/NA	UA/NA	UA/NA	HARVARD
6,001,718	0	4,417	5,631	255,967	3,090	33,131	28,538	HAWAII
4,531,918	680,971	2,105	2,940	5,399	526	4,705	5,689	HOUSTON
3,671,886	540	16	17,975	18	156,974	25,073	6,163	HOWARD
3,903,703	762,943	232	34,464	UA/NA	816,351	2,078	4,441	ILLINOIS, CHICAGO
9,302,770	UA/NA	12,484	34,994	658,905	90,570	152,511	21,124	ILLINOIS, URBANA

UA/NA – Unavailable or Not Applicable

ARL Library Data Tables, 2004-05
Collections

(Survey Question #) INSTITUTION	Notes	Volumes in Library (1)	Volumes Added (Gross) (1.b.i)	Volumes Added (Net) (1.b)	Monographs Purchased (Volumes) (2)	Current Serials Purchased (4.a)	Current Serials Received (4.b)	Current Serials Total (4)
INDIANA	L+	7,241,929	133,295	110,870	63,156	65,593	5,079	70,672
IOWA	GLM+	4,551,217	118,346	76,391	63,627	31,605	18,609	50,214
IOWA STATE	G+	2,444,263	48,171	27,593	29,029	16,698	13,152	29,850
JOHNS HOPKINS	GM+	3,648,821	54,597	42,567	30,874	43,014	30,858	73,872
KANSAS	LM+	4,121,573	83,469	81,928	38,123	27,721	23,271	50,992
KENT STATE	G+	2,715,986	50,327	48,303	10,752	11,176	2,181	13,357
KENTUCKY	GLM+	3,286,731	77,597	74,062	46,041	20,073	10,177	30,250
LAVAL	GLM+	2,731,341	103,857	73,214	32,600	21,438	4,951	26,389
LOUISIANA STATE	LM+	3,406,434	57,259	45,225	70,185	UA/NA	UA/NA	58,461
LOUISVILLE	GLM+	2,015,752	83,157	65,132	35,345	22,833	742	23,575
MCGILL	GLM+	3,568,651	68,311	52,856	40,229	42,794	6,639	49,433
MCMASTER	GM+	1,990,867	33,375	22,699	46,059	14,921	11,987	26,908
MANITOBA	GLM+	2,158,148	37,089	32,861	41,616	20,005	6,034	26,039
MARYLAND	G+	3,259,600	84,206	76,627	59,002	12,847	20,630	33,477
MASSACHUSETTS	G+	3,204,025	35,823	34,742	UA/NA	34,199	7,008	41,207
MIT	G+	2,782,406	44,256	40,462	19,207	12,570	10,468	23,038
MIAMI	GLM+	2,571,045	80,204	55,313	35,119	UA/NA	UA/NA	38,432
MICHIGAN	LM+	8,133,917	189,373	175,772	UA/NA	UA/NA	UA/NA	124,809
MICHIGAN STATE	GL+	4,830,861	60,188	42,270	34,427	21,326	16,506	37,832
MINNESOTA	BGLM+	6,587,430	148,474	124,725	34,075	31,098	12,205	43,303
MISSOURI	BGLM+	3,249,783	51,179	43,856	27,314	32,902	3,600	36,502
MONTREAL	GLM+	3,057,166	81,449	10,152	57,485	38,444	5,263	43,707
NEBRASKA	GL+	2,942,946	61,649	42,227	16,361	35,714	8,391	44,105
NEW MEXICO	LM+	2,666,970	56,162	39,155	23,806	11,891	1,446	13,337
NEW YORK	GLM+	4,939,724	108,920	82,136	56,881	41,121	12,903	54,024
NORTH CAROLINA	GLM+	5,710,686	119,474	109,250	57,520	28,862	24,582	53,444
NORTH CAROLINA STATE	G+	3,530,949	147,268	141,432	24,660	23,011	34,475	57,486
NORTHWESTERN	GLM+	4,603,824	89,205	58,786	39,181	UA/NA	UA/NA	41,693
NOTRE DAME	GL+	3,185,926	67,696	63,739	30,274	UA/NA	UA/NA	22,866
OHIO	BM+	2,599,791	57,126	49,280	42,950	16,346	3,582	19,928
OHIO STATE	GLM+	5,936,434	124,288	126,929	67,494	20,212	14,533	34,745
OKLAHOMA	GLM+	4,796,089	63,849	59,876	28,067	31,752	13,223	44,975
OKLAHOMA STATE	BGM+	2,624,680	78,575	72,996	23,897	UA/NA	UA/NA	61,507
OREGON	BGL+	2,643,035	33,110	30,392	20,782	13,033	5,793	18,826
PENNSYLVANIA	LM+	5,760,065	105,875	99,888	UA/NA	41,166	2,765	43,931
PENNSYLVANIA STATE	GLM+	5,031,196	81,008	55,857	59,347	54,155	14,290	68,445
PITTSBURGH	GLM+	4,786,175	153,774	145,896	UA/NA	UA/NA	UA/NA	43,441
PRINCETON	G+	6,495,597	127,750	122,413	88,070	62,921	1,066	63,987
PURDUE	G+	2,475,242	49,636	15,299	21,860	13,868	7,730	21,598
QUEEN'S	BLM+	2,437,168	31,930	26,299	21,149	20,947	974	21,921
RICE	G+	2,437,177	44,338	43,046	34,629	11,010	5,944	16,954
ROCHESTER	BGM+	3,564,683	53,544	45,504	20,753	UA/NA	UA/NA	26,218

UA/NA – Unavailable or Not Applicable
L - Includes Law library
M- Includes Medical library
+ - See Footnotes
B - Basis of volume count is bibliographic
G - Government documents not included in serials count

ARL Library Data Tables, 2004-05
Collections

Microform Units	Government Documents	Computer Files	Archives And Manus.	Carto-graphic Materials	Graphic Materials	Audio Materials	Video and Film	(Survey Question #) INSTITUTION
(6)	(7)	(8)	(9)	(10)	(11)	(12)	(13)	
4,978,205	405,323	59,184	36,336	678,972	3,976,930	281,866	53,583	INDIANA
7,131,140	665,618	22,650	15,779	418,745	229,393	34,133	22,060	IOWA
3,473,037	0	8,397	15,605	108,141	850,098	13,511	51,894	IOWA STATE
4,380,319	3,897	6,107	12,388	215,950	69,825	28,077	10,460	JOHNS HOPKINS
3,606,499	732,058	11,419	46,589	336,447	3,059,584	35,237	17,076	KANSAS
1,369,182	0	45,917	9,466	8,600	2,401	16,474	12,488	KENT STATE
6,347,361	1,125,174	7,353	25,592	257,418	2,089,208	48,232	40,399	KENTUCKY
1,312,518	0	5,220	0	314,700	216,647	19,094	23,231	LAVAL
6,697,824	613,533	8,139	23,389	324,798	300,012	22,223	4,352	LOUISIANA STATE
2,190,246	0	0	29,501	22,241	1,513,660	30,316	8,313	LOUISVILLE
1,793,955	668,033	6,099	3,408	241,192	258,163	48,400	36,428	MCGILL
1,597,371	0	1,143	12,383	146,371	10	29,813	1,470	MCMASTER
1,468,054	225,236	UA/NA	10,770	108,995	272,738	26,751	9,161	MANITOBA
5,736,084	185,837	7,086	35,858	223,364	343,487	271,943	37,212	MARYLAND
2,575,821	315,238	2,977	12,860	102,358	0	19,005	4,873	MASSACHUSETTS
2,391,503	UA/NA	20,643	18,114	122,172	452,092	28,473	4,522	MIT
3,924,529	0	4,602	17,389	48,978	12,218	50,168	12,209	MIAMI
UA/NA	UA/NA	UA/NA	UA/NA	UA/NA	UA/NA	UA/NA	UA/NA	MICHIGAN
6,518,542	0	23,206	2,600	274,978	1,817	63,445	2,633	MICHIGAN STATE
6,715,389	2,652,842	16,507	75,962	427,166	UA/NA	UA/NA	UA/NA	MINNESOTA
7,437,324	1,663,610	2,036	10,764	271,709	609	22,991	3,749	MISSOURI
1,516,331	0	8,736	0	13,155	132,445	39,574	13,232	MONTREAL
4,618,411	552,672	6,509	9,267	167,697	102,970	34,073	2,165	NEBRASKA
4,690,017	609,773	30,059	15,154	218,922	1,070,715	27,474	7,031	NEW MEXICO
6,332,917	450,565	5,525	29,808	24	1,363,536	92,663	38,296	NEW YORK
4,987,306	1,625,932	11,216	34,546	322,162	777,293	263,721	53,108	NORTH CAROLINA
5,413,654	0	156,722	11,576	40,519	116,525	3,913	19,553	NORTH CAROLINA STATE
4,405,025	7,808	13,558	13,550	206,126	610,543	58,108	25,136	NORTHWESTERN
3,831,274	UA/NA	5,946	UA/NA	19,336	UA/NA	20,522	8,771	NOTRE DAME
3,206,766	31,366	17,536	12,783	184,120	97,671	55,848	34,308	OHIO
5,834,511	6,287	11,580	29,854	223,524	2,294,014	48,901	UA/NA	OHIO STATE
4,316,009	18,480	4,509	17,205	155,704	1,731,533	4,248	8,185	OKLAHOMA
4,488,446	145,996	14,506	9,854	298,242	205,312	10,938	16,430	OKLAHOMA STATE
2,909,868	467,921	2,320	73,388	829,533	1,318,505	77,369	19,499	OREGON
4,048,727	UA/NA	8,988	12,254	134,998	741,635	63,773	12,171	PENNSYLVANIA
5,271,498	0	9,415	31,624	468,799	785,180	72,360	171,655	PENNSYLVANIA STATE
6,145,754	0	36,450	29,843	123,337	732,845	97,809	60,156	PITTSBURGH
6,424,007	123,609	10,602	32,654	330,614	190,577	87,237	4,620	PRINCETON
3,145,402	488,674	9,641	2,710	343,584	1,205	933	12,439	PURDUE
3,861,728	1,099,003	UA/NA	UA/NA	160,421	28,321	UA/NA	UA/NA	QUEEN'S
3,168,578	55,896	14,366	6,223	26,925	33,885	41,114	9,922	RICE
5,081,610	443,509	6,962	14,153	105,043	7,227	88,863	13,396	ROCHESTER

UA/NA – Unavailable or Not Applicable

ARL LIBRARY DATA TABLES, 2004-05
COLLECTIONS

(Survey Question #) INSTITUTION	Notes	Volumes in Library (1)	Volumes Added (Gross) (1.b.i)	Volumes Added (Net) (1.b)	Monographs Purchased (Volumes) (2)	Current Serials Purchased (4.a)	Current Serials Received (4.b)	Current Serials Total (4)
RUTGERS	GL+	4,169,347	76,198	61,809	23,331	47,060	3,169	50,229
SASKATCHEWAN	GLM+	1,976,606	57,844	26,024	21,137	19,632	6,249	25,881
SOUTH CAROLINA	BLM+	3,478,256	42,928	41,811	28,003	UA/NA	UA/NA	53,610
SOUTHERN CALIFORNIA	GLM+	3,921,704	63,817	55,790	34,422	29,916	7,893	37,809
SOUTHERN ILLINOIS	LM+	2,999,736	43,765	33,779	16,532	25,420	17,663	43,083
SUNY-ALBANY	G+	2,094,048	35,707	29,472	14,039	38,859	2,848	41,707
SUNY-BUFFALO	GLM+	3,390,583	48,556	30,547	15,436	22,145	12,820	34,965
SUNY-STONY BROOK	BM+	2,216,589	26,304	23,885	12,602	27,651	6,777	34,428
SYRACUSE	GL+	3,161,529	44,840	24,565	19,562	16,862	3,775	20,637
TEMPLE	BGLM+	3,016,007	44,143	43,546	28,418	18,717	8,262	26,979
TENNESSEE	GLM+	2,971,837	64,013	51,352	28,860	25,650	10,581	36,231
TEXAS	GL+	8,937,002	147,811	122,854	49,716	26,433	20,424	46,857
TEXAS A&M	G+	3,437,376	110,204	106,252	73,484	45,079	11,824	56,903
TEXAS TECH	LM+	2,465,796	64,585	55,103	38,588	25,193	17,667	42,860
TORONTO	GL+	10,342,574	269,669	241,024	142,023	61,357	6,318	67,675
TULANE	LM+	2,403,728	51,814	35,027	31,967	15,710	878	16,588
UTAH	BGLM+	3,230,854	67,556	44,944	UA/NA	UA/NA	UA/NA	40,446
VANDERBILT	GLM+	3,056,707	72,520	65,144	26,592	25,476	8,407	33,883
VIRGINIA	LM+	5,053,162	77,249	65,725	38,893	28,375	24,995	53,370
VIRGINIA TECH	G+	2,237,106	38,048	26,461	20,441	31,365	2,560	33,925
WASHINGTON	BGLM+	6,639,850	120,914	93,778	55,430	UA/NA	UA/NA	55,932
WASHINGTON STATE	G+	2,256,229	64,702	59,069	18,694	UA/NA	UA/NA	30,648
WASHINGTON U.-ST. LOUIS	GLM+	3,694,504	57,710	47,045	27,942	UA/NA	UA/NA	44,806
WATERLOO	G+	2,006,887	25,501	14,187	17,868	UA/NA	UA/NA	20,028
WAYNE STATE	G+	3,383,826	46,707	35,584	19,277	UA/NA	UA/NA	22,606
WESTERN ONTARIO	GL+	3,085,319	32,907	28,444	28,444	41,912	3,512	45,424
WISCONSIN	GLM+	7,911,834	124,409	104,737	54,891	44,710	15,747	60,457
YALE	GLM+	12,025,695	302,604	297,248	207,276	68,116	7,906	76,022
YORK	GL+	2,489,193	48,948	12,492	34,747	23,686	3,152	26,838
BOSTON PUBLIC	G+	7,488,088	125,567	44,329	65,295	UA/NA	UA/NA	7,301
CISTI	G+	8,141,477	33,585	25,892	2,667	UA/NA	UA/NA	8,702
CTR FOR RESEARCH LIBS	+	2,200,567	12,182	12,182	4,763	UA/NA	UA/NA	UA/NA
LIBRARY OF CONGRESS	BG+	26,683,947	364,956	361,384	127,527	37,125	62,875	100,000
NATL AGRICULTURAL LIB	G+	2,408,169	12,491	11,697	546	4,837	13,277	18,114
LIB & ARCHIVES CANADA	G+	9,052,702	354,808	UA/NA	7,586	1,771	34,000	35,771
NATL LIB OF MEDICINE	G+	2,512,114	41,318	29,529	15,944	19,142	4,018	23,160
NEW YORK PUBLIC	G+	8,828,221	117,305	117,296	0	56,187	26,049	82,236
NEW YORK STATE	+	2,632,305	33,558	33,475	10,173	7,573	6,660	14,233
SMITHSONIAN	+	1,551,323	21,692	20,513	4,125	2,657	2,233	4,890

UA/NA – Unavailable or Not Applicable
L - Includes Law library
M- Includes Medical library

+ - See Footnotes
B - Basis of volume count is bibliographic
G - Government documents not included in serials count

ARL Library Data Tables, 2004-05
Collections

Microform Units (6)	Government Documents (7)	Computer Files (8)	Archives and Manus. (9)	Carto-graphic Materials (10)	Graphic Materials (11)	Audio Materials (12)	Video and Film (13)	(Survey Question #) INSTITUTION
5,906,846	2,661,336	13,998	26,172	UA/NA	UA/NA	UA/NA	UA/NA	RUTGERS
3,069,886	412,224	5,071	268	72,790	19,386	17,011	2,032	SASKATCHEWAN
5,210,142	920,317	3,417	8,958	320,374	250,000	35,989	14,414	SOUTH CAROLINA
6,187,918	266,024	10,819	46,041	84,491	3,192,362	28,698	29,487	SOUTHERN CALIFORNIA
4,566,453	313,173	10,044	23,790	342,952	18,063	27,803	14,157	SOUTHERN ILLINOIS
2,860,553	0	2,221	6,934	24,714	55,451	7,961	7,150	SUNY-ALBANY
5,406,490	UA/NA	8,437	8,400	197,386	132,985	49,256	8,862	SUNY-BUFFALO
3,832,124	335,159	326	7,240	132,708	9,457	39,142	7,200	SUNY-STONY BROOK
7,334,236	306,460	1,984	31,530	276,369	416,691	410,892	19,934	SYRACUSE
3,147,750	87,621	3,610	35,640	120,666	10,034,528	23,499	5,365	TEMPLE
4,168,542	1,776	2,186	13,429	389,435	123,000	28,414	14,855	TENNESSEE
6,510,523	0	7,965	120,109	322,217	11,583,168	190,921	47,595	TEXAS
5,480,289	65,519	11,272	22,345	223,947	518,539	17,752	26,336	TEXAS A&M
2,591,561	4,222	13,855	17,332	157,219	1,395,929	14,213	23,214	TEXAS TECH
5,371,934	UA/NA	449,475	37,568	282,217	721,683	196,634	28,128	TORONTO
2,652,736	616,614	3,205	46,958	29,946	426,113	88,015	8,627	TULANE
3,608,147	885,549	7,420	32,090	250,738	1,395,723	47,161	30,139	UTAH
3,022,373	255,423	10,694	15,877	108,284	221,680	34,828	9,705	VANDERBILT
5,541,742	676,334	21,289	27,348	151,461	465,037	65,043	29,677	VIRGINIA
6,292,655	399,525	10,566	18,503	136,990	86,493	9,914	15,262	VIRGINIA TECH
7,588,429	0	12,749	68,799	364,258	1,219,846	73,554	28,308	WASHINGTON
4,224,734	UA/NA	UA/NA	18,874	2,119	366,419	13,928	35,072	WASHINGTON STATE
3,355,312	0	9,051	14,169	119,161	119,240	39,203	34,993	WASHINGTON U.-ST. LOUIS
1,707,697	0	5,495	5,021	150,983	UA/NA	1,016	162	WATERLOO
3,847,036	388,182	5,138	74,000	54,910	2,158,539	46,156	28,023	WAYNE STATE
3,883,392	0	297	11,437	15,251	1,600,000	54,908	1,715	WESTERN ONTARIO
6,248,997	909,069	11,678	155,457	573,450	6,174,957	332,641	20,676	WISCONSIN
10,037,556	0	1,296	76,999	459,942	2,280,109	231,970	16,710	YALE
4,101,547	0	UA/NA	9,951	113,219	8,647	32,729	21,453	YORK
6,741,213	1,759,457	130	UA/NA	UA/NA	UA/NA	57,033	74,213	BOSTON PUBLIC
5,000,000	0	22,409	UA/NA	UA/NA	UA/NA	UA/NA	806	CISTI
1,822,504	0	781	0	0	0	0	0	CTR FOR RESEARCH LIBS
14,196,016	UA/NA	124,578	58,952	5,235,795	14,298,934	2,824,294	982,099	LIBRARY OF CONGRESS
1,087,138	0	1,927	19,165	4,960	1,125	462	3,536	NATL AGRICULTURAL LIB
7,966,712	3,697,513	UA/NA	555,590	2,936,806	24,610,855	164,427	188,821	LIB & ARCHIVES CANADA
598,503	0	2,469	3,035	0	63,255	UA/NA	77,344	NATL LIB OF MEDICINE
6,806,099	0	UA/NA	64,979	432,106	4,373,415	568,818	59,738	NEW YORK PUBLIC
8,061,040	UA/NA	6,977	27,715	208,305	70,304	55	84	NEW YORK STATE
190,237	UA/NA	1,479	2,124	118	2,013	212	298	SMITHSONIAN

UA/NA – Unavailable or Not Applicable

ARL LIBRARY DATA TABLES, 2004-05
SUMMARY DATA: COLLECTIONS

	Volumes in Library	Volumes Added (Gross)	Volumes Added (Net)	Monographs Purchased (Volumes)	Current Serials Purchased	Current Serials Received	Current Serials Total
(Survey Question #)	(1)	(1.b.i)	(1.b)	(2)	(4.a)	(4.b)	(4)
UNIVERSITY LIBRARIES:							
Median	3,230,854	64,013	54,437	32,712	24,654	8,327	37,668
High	15,555,533	429,344	297,248	249,680	68,194	34,475	124,809
Low	1,559,078	19,453	3,693	3,689	7,518	315	11,317
Totals	458,151,678	9,255,284	7,523,034	4,109,247	2,493,618	878,218	4,658,493
Number of Libraries Responding	113	113	113	101	88	88	113
NONUNIVERSITY LIBRARIES:							
Median	5,060,197	37,452	29,529	6,175	7,573	13,277	18,114
Totals	71,498,913	1,117,462	656,297	238,626	129,292	149,112	294,407
Number of Libraries Responding	10	10	9	10	7	7	9
GRAND TOTALS	529,650,591	10,372,746	8,179,331	4,347,873	2,622,910	1,027,330	4,952,900

ARL Library Data Tables, 2004-05
Summary Data: Collections

Microform Units	Government Documents	Computer Files	Archives and Manus.	Carto-graphic Materials	Graphic Materials	Audio Materials	Video and Film	
(6)	(7)	(8)	(9)	(10)	(11)	(12)	(13)	(Survey Question #)
								UNIVERSITY LIBRARIES:
4,227,615	193,534	8,988	17,269	173,621	242,470	39,142	15,251	Median
10,037,556	2,661,336	449,475	232,205	3,895,944	11,583,168	410,892	171,655	High
1,312,518	0	0	0	18	0	933	162	Low
511,328,095	37,303,170	1,706,018	2,841,085	28,272,673	91,323,151	6,732,498	2,179,695	Totals
112	99	107	108	109	106	107	106	Number of Libraries Responding
								NONUNIVERSITY LIBRARIES:
5,870,607	0	2,198	23,440	106,633	66,780	28,748	31,637	Median
52,469,462	5,456,970	160,750	731,560	8,818,090	43,419,901	3,615,301	1,386,939	Totals
10	7	8	8	8	8	8	10	Number of Libraries Responding
563,797,557	42,760,140	1,866,768	3,572,645	37,090,763	134,743,052	10,347,799	3,566,634	**GRAND TOTALS**

ARL Library Data Tables, 2004-05
Expenditures

(Survey Question #) INSTITUTION	Notes	Monographs (15.a)	Current Serials (15.b)	Other Library Materials (15.c)	Misc. Materials (15.d)	Total Library Materials (15)	Contract Binding (16)
ALABAMA	BGLM+	1,712,729	4,515,297	7,704	38,576	6,274,306	79,546
ALBERTA	GLM+	4,295,457	7,837,627	0	555,942	12,689,026	94,259
ARIZONA	BGLM+	2,495,290	7,743,061	712,127	339,124	11,289,602	98,696
ARIZONA STATE	GL+	2,150,937	6,758,072	808,301	158,210	9,875,520	173,042
AUBURN	G+	891,275	3,440,754	860,518	68,849	5,261,396	76,811
BOSTON	GLM+	887,337	6,184,327	853,124	3,178	7,927,966	186,662
BOSTON COLLEGE	BGL+	1,618,129	5,824,029	49,866	291,453	7,783,477	141,410
BRIGHAM YOUNG	BGL+	2,106,921	8,011,243	89,439	93,992	10,301,595	204,250
BRITISH COLUMBIA	GLM+	1,969,804	8,431,982	613,897	778,493	11,794,175	153,397
BROWN	GM+	2,335,426	5,040,773	334,082	354,560	8,064,841	226,081
CALIFORNIA, BERKELEY	L+	7,385,144	7,947,095	200,132	UA/NA	15,532,371	902,525
CALIFORNIA, DAVIS	GLM+	2,033,057	3,461,034	714,450	209,143	6,417,684	153,480
CALIFORNIA, IRVINE	BGM+	1,784,841	4,757,232	383,936	UA/NA	6,926,009	83,694
CALIFORNIA, LOS ANGELES	GLM+	4,620,063	6,685,201	75,810	511,432	11,892,506	509,591
CALIFORNIA, RIVERSIDE	BG+	1,558,422	3,799,154	77,017	10,190	5,444,783	243,746
CALIFORNIA, SAN DIEGO	GM+	2,640,757	5,369,676	314,489	0	8,324,922	307,740
CALIFORNIA, SANTA BARBARA	+	2,076,688	3,635,558	UA/NA	0	5,712,246	184,020
CASE WESTERN RESERVE	GLM+	772,147	5,443,602	82,774	485,728	6,784,251	93,418
CHICAGO	BGLM+	4,360,625	8,897,582	231,396	0	13,489,603	328,116
CINCINNATI	GLM+	2,901,075	6,178,530	547,326	447,539	10,074,470	119,446
COLORADO	BG+	1,584,992	6,398,185	345,961	472,824	8,801,962	95,658
COLORADO STATE	G+	1,587,712	4,279,800	36,432	333,600	6,237,544	116,616
COLUMBIA	LM+	6,035,280	11,360,265	1,062,880	701,472	19,159,897	659,284
CONNECTICUT	GLM+	1,091,820	6,564,118	137,391	407,034	8,200,363	244,353
CORNELL	GL+	5,234,477	7,500,212	813,693	226,107	13,774,489	184,047
DARTMOUTH	GM+	1,712,646	5,466,183	59,516	0	7,238,345	118,675
DELAWARE	G+	2,149,791	4,934,128	158,422	122,634	7,364,975	82,341
DUKE	BLM+	3,020,512	8,286,199	115,759	857,099	12,279,569	170,411
EMORY	LM+	4,172,594	5,405,397	3,308,431	483,839	13,370,261	141,657
FLORIDA	BGLM+	2,232,834	7,867,232	598,242	UA/NA	10,698,308	298,410
FLORIDA STATE	BLM+	1,319,330	4,336,516	208,013	1,038,522	6,902,381	70,013
GEORGE WASHINGTON	BGLM+	3,007,868	5,102,310	117,356	945,433	9,172,967	143,273
GEORGETOWN	BGLM	1,627,048	6,743,739	921,656	78,828	9,371,271	126,496
GEORGIA	GL+	2,151,225	7,465,945	1,002,956	393,870	11,013,996	317,280
GEORGIA TECH	G+	490,380	3,614,169	1,189,458	0	5,294,007	52,807
GUELPH	G+	1,036,737	3,047,922	378,905	257,827	4,721,391	25,106
HARVARD	GLM+	9,595,629	10,257,953	7,289,545	426,696	27,569,823	1,060,533
HAWAII	G+	1,243,398	4,573,379	93,588	0	5,910,365	200,000
HOUSTON	BGL+	1,331,054	5,670,011	402,932	277,829	7,681,826	112,499
HOWARD	BLM	239,204	3,501,658	127,630	17,500	3,885,992	57,500
ILLINOIS, CHICAGO	GM+	1,097,522	6,087,267	545,047	UA/NA	7,729,836	74,268
ILLINOIS, URBANA	GL+	2,644,792	9,520,111	123,870	765,054	13,053,827	309,333

UA/NA - Unavailable or Not Applicable
L - Includes Law library
M- Includes Medical library

+ - See Footnotes
B - Basis of volume count is bibliographic
G - Government documents not included in serials count

ARL Library Data Tables, 2004-05
Expenditures

Salaries & Wages: Professional	Salaries & Wages: Support	Salaries & Wages: Student Asst.	Total Salaries & Wages	Other Operating Expenditures	Total Library Expenditures	
(17.a)	(17.b)	(17.c)	(17)	(19)	(20)	(Survey Question #) INSTITUTION
3,035,555	1,795,270	683,271	5,514,096	1,719,751	13,587,699	ALABAMA
4,595,537	7,334,649	801,672	12,731,858	1,254,525	26,769,667	ALBERTA
5,631,631	5,126,394	905,679	11,663,704	4,290,717	27,342,719	ARIZONA
4,973,188	5,959,272	708,946	11,641,406	2,891,586	24,581,554	ARIZONA STATE
2,722,767	1,405,555	455,005	4,583,327	2,330,330	12,251,864	AUBURN
4,781,994	3,106,270	462,296	8,350,560	2,068,658	18,533,846	BOSTON
4,630,773	2,580,155	733,302	7,944,230	1,011,046	16,880,163	BOSTON COLLEGE
6,312,687	1,729,858	2,916,269	10,958,814	3,507,022	24,971,681	BRIGHAM YOUNG
5,951,906	5,847,115	1,035,633	12,834,654	2,436,053	27,218,279	BRITISH COLUMBIA
4,107,988	3,912,735	418,765	8,439,488	1,748,566	18,478,976	BROWN
15,610,774	6,229,184	4,535,301	26,375,259	8,581,987	51,392,142	CALIFORNIA, BERKELEY
3,985,889	4,644,281	820,187	9,450,357	1,672,602	17,694,123	CALIFORNIA, DAVIS
3,044,410	4,846,975	629,185	8,520,570	2,047,034	17,577,307	CALIFORNIA, IRVINE
11,177,359	11,523,674	2,781,822	25,482,855	8,127,524	46,012,476	CALIFORNIA, LOS ANGELES
2,867,228	2,252,086	397,107	5,516,421	1,385,340	12,590,290	CALIFORNIA, RIVERSIDE
5,778,041	6,333,450	1,377,548	13,489,039	4,673,262	26,794,963	CALIFORNIA, SAN DIEGO
2,918,115	4,150,349	957,151	8,025,615	3,706,602	17,628,483	CALIFORNIA, SANTA BARBARA
3,376,439	1,709,719	525,527	5,611,685	2,107,843	14,597,197	CASE WESTERN RESERVE
4,107,460	5,874,748	1,197,672	11,179,880	2,745,954	27,743,553	CHICAGO
6,699,721	2,300,486	670,007	9,670,214	2,128,509	21,992,639	CINCINNATI
2,735,300	4,393,761	811,841	7,940,902	1,710,577	18,549,099	COLORADO
2,638,116	2,150,717	602,203	5,391,036	1,856,857	13,602,053	COLORADO STATE
14,024,076	7,778,330	1,723,316	23,525,722	7,106,593	50,451,496	COLUMBIA
6,910,329	5,094,372	831,216	12,835,917	1,699,787	22,980,420	CONNECTICUT
7,748,513	11,725,634	892,361	20,366,508	7,216,615	41,541,659	CORNELL
3,584,644	3,427,201	425,505	7,437,350	1,941,533	16,735,903	DARTMOUTH
3,698,748	3,019,456	399,207	7,117,411	1,329,026	15,893,753	DELAWARE
8,954,927	4,669,347	421,631	14,045,905	3,369,857	29,865,742	DUKE
5,903,599	5,319,407	718,041	11,941,047	4,498,938	29,951,903	EMORY
6,063,533	5,742,417	781,469	12,587,419	2,631,993	26,216,130	FLORIDA
2,774,384	2,262,404	428,474	5,465,262	2,947,469	15,385,125	FLORIDA STATE
4,740,461	3,564,095	613,563	8,918,119	1,975,812	20,210,171	GEORGE WASHINGTON
5,107,944	4,526,340	889,018	10,523,302	2,593,397	22,614,466	GEORGETOWN
3,264,824	5,130,941	762,909	9,158,674	2,189,915	22,679,865	GEORGIA
3,036,673	1,278,317	99,697	4,414,687	1,145,210	10,906,711	GEORGIA TECH
2,091,560	2,482,710	218,238	4,792,509	1,530,113	11,069,118	GUELPH
35,504,369	20,507,207	1,597,489	57,609,065	16,656,008	102,895,429	HARVARD
3,764,270	2,725,851	771,877	7,261,998	1,213,637	14,586,000	HAWAII
3,036,889	2,129,310	613,291	5,779,490	2,745,186	16,319,001	HOUSTON
2,771,164	2,190,294	332,334	5,293,792	775,653	10,012,937	HOWARD
4,149,601	3,498,586	431,784	8,079,971	1,196,888	17,080,963	ILLINOIS, CHICAGO
9,850,006	6,901,356	1,549,766	18,301,128	3,776,614	35,440,902	ILLINOIS, URBANA

UA/NA - Unavailable or Not Applicable

ARL Library Data Tables, 2004-05
Expenditures

(Survey Question #) INSTITUTION	Notes	Monographs (15.a)	Current Serials (15.b)	Other Library Materials (15.c)	Misc. Materials (15.d)	Total Library Materials (15)	Contract Binding (16)
INDIANA	L+	3,966,347	6,147,597	2,319,266	240,960	12,674,170	317,484
IOWA	GLM+	2,415,944	7,017,808	1,884,546	548,075	11,866,373	177,867
IOWA STATE	G+	1,794,752	6,337,530	84,266	383,238	8,599,786	153,388
JOHNS HOPKINS	GM+	1,545,873	9,881,835	399,383	452,647	12,279,738	100,771
KANSAS	LM+	1,811,413	5,634,506	278,650	608,636	8,333,205	133,219
KENT STATE	G+	622,190	2,738,556	71,542	0	3,432,288	85,689
KENTUCKY	GLM+	1,532,372	6,854,003	174,647	852,405	9,413,427	189,816
LAVAL	GLM+	1,546,700	5,994,235	121,791	140,320	7,803,045	35,788
LOUISIANA STATE	LM+	2,258,247	4,386,085	288,107	0	6,932,439	68,548
LOUISVILLE	GLM+	1,312,102	7,078,692	224,818	25,484	8,641,096	100,571
MCGILL	GLM+	2,756,401	6,518,215	1,233,379	868,638	11,376,632	169,965
MCMASTER	GM+	1,218,069	4,312,240	312,362	142,244	5,984,915	152,335
MANITOBA	GLM+	1,405,048	4,240,389	201,604	516,711	6,363,752	92,804
MARYLAND	G+	1,665,420	3,996,874	313,577	2,439,294	8,415,165	129,818
MASSACHUSETTS	G+	910,277	4,094,358	228,477	45,094	5,278,206	96,800
MIT	G+	1,282,263	5,802,131	0	55,658	7,140,052	210,385
MIAMI	GLM+	2,242,161	6,944,650	1,123,049	333,848	10,643,708	151,549
MICHIGAN	LM+	UA/NA	UA/NA	UA/NA	UA/NA	20,309,007	UA/NA
MICHIGAN STATE	GL+	2,053,447	7,090,194	210,287	133,474	9,487,402	225,978
MINNESOTA	BGLM+	2,411,411	9,262,586	742,921	142,771	12,559,689	335,922
MISSOURI	BGLM+	964,293	4,636,091	173,054	60,001	5,833,439	143,757
MONTREAL	GLM+	2,394,073	6,862,460	84,615	176,617	9,517,765	217,092
NEBRASKA	GL+	900,763	5,180,965	418,765	72,368	6,572,861	159,180
NEW MEXICO	LM+	850,138	4,447,558	162,159	357,182	5,817,037	140,337
NEW YORK	GLM+	3,169,855	9,715,687	1,456,973	112,035	14,454,550	630,776
NORTH CAROLINA	GLM+	2,786,918	7,608,439	2,319,282	52,275	12,766,914	265,717
NORTH CAROLINA STATE	G+	1,732,653	5,875,549	523,260	742,233	8,873,695	184,471
NORTHWESTERN	GLM+	3,078,747	7,537,298	481,888	438,392	11,536,325	240,897
NOTRE DAME	GL+	2,941,063	5,012,370	423,954	285,281	8,662,668	89,421
OHIO	BM+	1,111,982	3,788,272	440,331	50,447	5,391,032	54,188
OHIO STATE	GLM+	2,288,345	4,738,482	1,402,531	450,702	8,880,060	252,445
OKLAHOMA	GLM+	3,291,713	5,375,805	2,473,125	128,255	11,268,898	192,494
OKLAHOMA STATE	BGM+	1,410,759	4,071,623	1,060,625	147,024	6,690,031	70,262
OREGON	BGL+	1,335,830	3,367,229	UA/NA	631,192	5,334,251	96,944
PENNSYLVANIA	LM+	3,320,341	9,222,287	557,857	UA/NA	13,100,485	347,961
PENNSYLVANIA STATE	GLM+	3,230,617	5,933,378	8,181,691	241,841	17,587,527	500,639
PITTSBURGH	GLM+	3,063,051	8,182,638	1,359,227	465,572	13,070,488	271,487
PRINCETON	G+	5,693,652	6,782,562	1,487,710	266,741	14,230,665	580,490
PURDUE	G+	1,498,189	6,687,528	1,356,301	0	9,542,018	97,845
QUEEN'S	BLM+	1,120,089	5,141,295	79,300	427,205	6,767,889	94,028
RICE	G+	3,308,947	4,767,337	476,014	326,134	8,878,432	46,581
ROCHESTER	BGM+	1,011,000	5,024,194	342,484	44,320	6,421,998	116,304

UA/NA – Unavailable or Not Applicable
L - Includes Law library
M- Includes Medical library
+ - See Footnotes
B - Basis of volume count is bibliographic
G - Government documents not included in serials count

ARL LIBRARY DATA TABLES, 2004-05
EXPENDITURES

Salaries & Wages: Professional	Salaries & Wages: Support	Salaries & Wages: Student Asst.	Total Salaries & Wages	Other Operating Expenditures	Total Library Expenditures	
(17.a)	(17.b)	(17.c)	(17)	(19)	(20)	(Survey Question #) INSTITUTION
7,593,176	4,630,229	1,692,405	13,915,810	3,817,639	30,725,103	INDIANA
4,900,121	5,034,171	993,568	10,927,860	1,676,573	24,648,673	IOWA
2,857,855	3,418,989	519,650	6,796,494	1,147,520	16,697,188	IOWA STATE
6,101,258	4,922,351	475,859	11,499,468	4,685,790	28,565,767	JOHNS HOPKINS
4,839,771	2,876,484	909,950	8,626,205	2,219,190	19,311,819	KANSAS
2,582,024	1,394,443	436,867	4,413,334	1,504,737	9,436,048	KENT STATE
4,637,258	2,661,704	830,822	8,129,784	2,074,958	19,807,985	KENTUCKY
3,056,772	4,794,659	203,031	8,054,461	482,730	16,376,024	LAVAL
2,887,021	1,966,601	652,674	5,506,296	1,205,455	13,712,738	LOUISIANA STATE
2,976,689	2,291,800	551,849	5,820,338	2,050,019	16,612,024	LOUISVILLE
4,158,017	4,728,779	827,322	9,714,118	3,719,813	24,980,527	MCGILL
1,807,771	3,101,974	290,724	5,200,469	714,869	12,052,587	MCMASTER
3,601,348	4,053,541	567,012	8,221,900	1,317,604	15,996,061	MANITOBA
6,850,799	3,009,096	754,858	10,614,753	2,681,756	21,841,492	MARYLAND
3,366,532	2,826,691	472,714	6,665,937	1,374,144	13,415,087	MASSACHUSETTS
7,022,558	3,672,465	313,401	11,008,424	2,304,151	20,663,012	MIT
4,196,898	2,486,376	836,605	7,519,879	2,247,638	20,562,774	MIAMI
UA/NA	UA/NA	UA/NA	UA/NA	UA/NA	47,113,239	MICHIGAN
4,369,713	4,722,962	1,391,088	10,483,763	2,614,114	22,811,257	MICHIGAN STATE
6,225,080	8,292,532	1,546,123	16,063,735	5,907,137	34,866,483	MINNESOTA
3,004,354	2,674,852	356,120	6,035,326	2,239,546	14,252,068	MISSOURI
5,664,771	8,089,661	44,135	13,798,567	1,169,143	24,702,567	MONTREAL
2,666,983	2,813,826	653,652	6,134,461	2,150,177	15,016,679	NEBRASKA
4,871,365	5,531,879	517,619	10,920,863	2,315,100	19,193,337	NEW MEXICO
7,826,698	8,397,033	1,129,461	17,353,192	5,746,250	38,184,768	NEW YORK
8,798,396	5,737,208	978,762	15,514,366	3,942,079	32,489,076	NORTH CAROLINA
6,213,658	2,803,310	846,535	9,863,503	4,627,288	23,548,957	NORTH CAROLINA STATE
6,969,784	4,469,467	675,390	12,114,641	2,599,170	26,491,033	NORTHWESTERN
3,735,117	4,541,138	393,392	8,669,647	1,827,568	19,249,304	NOTRE DAME
2,835,718	2,331,761	589,329	5,756,808	2,716,296	13,918,324	OHIO
7,247,429	5,265,733	1,519,907	14,033,069	3,789,213	26,954,787	OHIO STATE
2,813,955	1,940,951	852,599	5,607,505	1,500,869	18,569,766	OKLAHOMA
2,996,223	1,285,767	637,169	4,919,159	2,081,186	13,760,638	OKLAHOMA STATE
3,052,065	2,962,049	1,032,278	7,046,392	894,208	13,371,795	OREGON
6,864,940	6,720,622	1,118,655	14,704,217	3,997,747	32,150,410	PENNSYLVANIA
7,629,265	13,829,770	895,577	22,354,612	5,806,273	46,249,051	PENNSYLVANIA STATE
5,932,084	4,460,971	852,381	11,245,436	3,297,410	27,884,821	PITTSBURGH
8,744,376	7,274,616	881,756	16,900,748	4,001,682	35,713,585	PRINCETON
4,285,031	4,442,726	650,296	9,378,053	2,512,384	21,530,300	PURDUE
1,867,154	3,559,378	257,124	5,683,656	1,807,127	14,352,701	QUEEN'S
3,286,368	1,418,071	150,848	4,855,287	1,432,729	15,213,029	RICE
5,457,678	2,173,556	453,704	8,084,938	1,881,598	16,504,838	ROCHESTER

UA/NA – Unavailable or Not Applicable

ARL LIBRARY DATA TABLES, 2004-05
EXPENDITURES

(Survey Question #) INSTITUTION	Notes	Monographs (15.a)	Current Serials (15.b)	Other Library Materials (15.c)	Misc. Materials (15.d)	Total Library Materials (15)	Contract Binding (16)
RUTGERS	GL+	1,351,966	7,364,328	811,698	499,444	10,027,436	144,475
SASKATCHEWAN	GLM+	1,697,999	4,528,334	UA/NA	156,185	6,382,518	75,071
SOUTH CAROLINA	BLM+	1,236,288	4,941,208	126,642	59,944	6,364,082	80,375
SOUTHERN CALIFORNIA	GLM+	2,024,772	6,941,777	2,797,125	93,716	11,857,390	135,106
SOUTHERN ILLINOIS	LM+	962,335	5,488,465	256,312	366,746	7,073,858	140,737
SUNY-ALBANY	G+	888,311	3,886,662	16,892	330,456	5,122,321	82,428
SUNY-BUFFALO	GLM+	1,043,075	5,873,962	302,530	171,524	7,391,091	117,507
SUNY-STONY BROOK	BM+	629,154	4,757,927	150,260	679,883	6,217,224	54,395
SYRACUSE	GL+	988,995	4,136,091	UA/NA	346,005	5,351,232	61,337
TEMPLE	BGLM+	1,206,321	4,545,975	1,540,544	2,749	7,295,589	92,162
TENNESSEE	GLM+	1,907,902	7,940,141	102,448	195,522	10,146,013	129,140
TEXAS	GL+	3,684,604	7,862,167	1,602,032	591	13,149,394	182,399
TEXAS A&M	G+	5,202,466	8,678,296	332,991	256,139	14,469,892	159,945
TEXAS TECH	LM+	2,213,875	6,324,203	154,185	577,013	9,269,276	100,136
TORONTO	GL+	6,929,869	9,446,724	3,439,797	UA/NA	19,816,389	356,844
TULANE	LM+	1,536,290	4,662,815	16,662	310,015	6,525,782	124,223
UTAH	BGLM+	1,858,489	5,422,996	77,921	121,629	7,481,035	171,358
VANDERBILT	GLM+	1,658,666	8,088,649	122,402	97,778	9,967,495	123,743
VIRGINIA	LM+	2,128,109	6,464,325	1,237,345	252,823	10,082,602	138,242
VIRGINIA TECH	G+	1,241,043	4,253,860	29,541	403,321	5,927,765	77,355
WASHINGTON	BGLM+	3,078,929	9,086,013	108,649	158,511	12,432,102	298,494
WASHINGTON STATE	G+	897,590	4,628,803	51,571	183,084	5,761,048	73,662
WASHINGTON U.-ST. LOUIS	GLM+	1,397,670	7,203,389	917,204	451,452	9,969,715	108,785
WATERLOO	G+	878,710	4,129,732	977,416	208,810	6,194,668	54,853
WAYNE STATE	G+	1,349,418	5,991,513	UA/NA	324,715	7,665,646	78,472
WESTERN ONTARIO	GL+	1,767,990	7,101,517	UA/NA	UA/NA	8,869,507	121,552
WISCONSIN	GLM+	2,532,640	6,060,813	1,939,539	405,673	10,938,665	287,556
YALE	GLM+	11,104,116	7,891,785	12,199,023	0	31,194,924	634,726
YORK	GL+	2,350,167	4,835,964	400,093	0	7,586,224	144,480
BOSTON PUBLIC	G+	3,322,083	1,372,708	1,221,277	27,663	5,943,731	16,620
CISTI	G+	429,300	8,563,632	703,995	54,580	9,751,507	UA/NA
CTR FOR RESEARCH LIBS	+	43,316	193,109	581,110	18,207	835,742	0
LIBRARY OF CONGRESS	BG+	3,721,993	3,160,280	4,720,495	903,044	12,505,812	1,432,595
NATL AGRICULTURAL LIB	G+	40,167	3,500,131	UA/NA	UA/NA	3,540,298	2,500
LIB & ARCHIVES CANADA	G+	478,359	695,255	207,911	19,183	1,400,708	5,344
NATL LIB OF MEDICINE	G+	1,243,228	6,883,412	128,802	9,000	8,264,442	154,455
NEW YORK PUBLIC	G+	4,529,410	5,211,438	2,131,053	0	11,871,901	569,675
NEW YORK STATE	+	781,000	2,541,506	0	0	3,322,506	26,035
SMITHSONIAN	+	1,601,833	1,061,550	4,464	102,884	2,770,731	5,426

UA/NA – Unavailable or Not Applicable
L - Includes Law library
M- Includes Medical library
+ - See Footnotes
B - Basis of volume count is bibliographic
G - Government documents not included in serials count

ARL LIBRARY DATA TABLES, 2004-05
EXPENDITURES

Salaries & Wages: Professional	Salaries & Wages: Support	Salaries & Wages: Student Asst.	Total Salaries & Wages	Other Operating Expenditures	Total Library Expenditures	
(17.a)	(17.b)	(17.c)	(17)	(19)	(20)	(Survey Question #) INSTITUTION
6,892,407	9,636,093	1,687,992	18,216,492	1,867,124	30,255,527	RUTGERS
2,477,399	2,492,160	269,720	5,239,279	828,024	12,524,891	SASKATCHEWAN
3,684,233	3,243,030	524,749	7,452,012	4,453,321	18,349,790	SOUTH CAROLINA
8,929,536	4,794,452	1,918,693	15,642,681	5,859,726	33,494,903	SOUTHERN CALIFORNIA
2,788,525	2,621,377	948,939	6,358,841	1,015,932	14,589,368	SOUTHERN ILLINOIS
3,379,760	2,215,516	593,862	6,189,138	854,575	12,248,462	SUNY-ALBANY
7,382,883	1,538,398	899,744	9,821,025	1,430,453	18,760,076	SUNY-BUFFALO
4,729,093	888,532	555,044	6,172,669	586,568	13,030,856	SUNY-STONY BROOK
3,467,616	3,666,410	289,707	7,423,733	1,158,783	13,995,085	SYRACUSE
3,923,442	2,001,608	638,743	6,563,793	1,321,479	15,273,023	TEMPLE
5,139,389	3,664,883	676,374	9,480,646	2,115,532	21,871,331	TENNESSEE
8,232,968	10,054,509	1,585,578	19,873,055	5,737,925	38,942,773	TEXAS
6,430,141	3,426,453	966,481	10,823,075	3,193,909	28,646,821	TEXAS A&M
4,256,689	2,831,624	1,254,801	8,343,114	3,020,954	20,733,480	TEXAS TECH
10,224,641	13,310,541	3,246,267	26,781,448	6,144,620	53,099,301	TORONTO
2,881,538	2,255,750	353,296	5,490,584	1,121,809	13,262,398	TULANE
3,885,113	5,850,574	1,415,645	11,151,332	4,029,797	22,833,522	UTAH
5,537,670	3,148,852	341,995	9,028,517	2,276,419	21,396,174	VANDERBILT
6,960,365	9,182,416	768,887	16,911,668	5,861,763	32,994,275	VIRGINIA
2,036,073	2,515,109	425,167	4,976,349	1,527,510	12,508,979	VIRGINIA TECH
9,324,381	6,152,007	2,035,156	17,511,544	5,519,261	35,761,401	WASHINGTON
2,610,506	2,902,513	467,158	5,980,177	2,120,059	13,934,946	WASHINGTON STATE
6,854,832	3,712,385	720,175	11,287,392	15,118,293	36,484,185	WASHINGTON U.-ST. LOUIS
2,013,201	2,797,782	596,986	5,407,968	632,989	12,290,479	WATERLOO
4,926,371	2,860,312	1,205,328	8,992,011	5,525,405	22,261,534	WAYNE STATE
3,193,702	2,851,784	400,222	6,445,707	604,219	16,040,986	WESTERN ONTARIO
12,698,305	6,090,137	2,822,926	21,611,368	6,645,300	39,482,889	WISCONSIN
16,170,764	11,755,815	1,290,211	29,216,790	7,742,895	68,789,335	YALE
3,816,767	3,814,494	839,814	8,471,075	3,824,183	20,025,962	YORK
10,625,368	10,822,691	557,365	22,005,424	5,987,968	33,953,743	BOSTON PUBLIC
UA/NA	UA/NA	UA/NA	15,497,263	10,654,827	35,895,528	CISTI
1,344,314	297,628	283,490	1,925,432	951,798	3,712,972	CTR FOR RESEARCH LIBS
307,061,000	UA/NA	UA/NA	307,061,000	283,421,594	604,421,001	LIBRARY OF CONGRESS
12,143,058	3,578,427	55,215	15,776,700	5,684,981	25,004,479	NATL AGRICULTURAL LIB
32,683,169	17,052,686	307,224	50,043,079	26,482,193	77,931,325	LIB & ARCHIVES CANADA
15,180,060	3,847,065	332,267	19,359,392	31,590,103	59,368,392	NATL LIB OF MEDICINE
UA/NA	UA/NA	1,377,578	27,786,455	9,943,767	50,171,798	NEW YORK PUBLIC
UA/NA	UA/NA	UA/NA	6,947,288	1,002,950	11,298,779	NEW YORK STATE
4,621,533	1,980,314	UA/NA	6,601,847	2,403,921	11,781,925	SMITHSONIAN

UA/NA – Unavailable or Not Applicable

ARL LIBRARY DATA TABLES, 2004-05
SUMMARY DATA: EXPENDITURES

	Monographs	Current Serials	Other Library Materials	Misc. Materials	Total Library Materials	Contract Binding
(Survey Question #)	(15.a)	(15.b)	(15.c)	(15.d)	(15)	(16)
UNIVERSITY LIBRARIES:						
Median	1,789,797	5,904,464	344,223	252,823	8,662,668	141,074
High	11,104,116	11,360,265	12,199,023	2,439,294	31,194,924	1,060,533
Low	239,204	2,738,556	0	0	3,432,288	25,106
Totals	256,204,183	680,774,493	91,178,710	32,211,618	1,080,558,152	21,092,060
Number of Libraries Responding	112	112	106	105	113	112
NONUNIVERSITY LIBRARIES:						
Median	1,012,114	2,850,893	581,110	19,183	4,742,015	16,620
Totals	16,190,689	33,183,021	9,699,108	1,134,561	60,207,378	2,212,650
Number of Libraries Responding	10	10	9	9	10	9
GRAND TOTALS	272,394,872	713,957,515	100,877,818	33,346,178	1,140,765,530	23,304,711

ARL LIBRARY DATA TABLES, 2004-05
SUMMARY DATA: EXPENDITURES

Salaries & Wages: Professional	Salaries & Wages: Support	Salaries & Wages: Student Asst.	Total Salaries & Wages	Other Operating Expenditures	Total Library Expenditures	
(17.a)	(17.b)	(17.c)	(17)	(19)	(20)	(Survey Question #)
						UNIVERSITY LIBRARIES:
4,482,625	3,665,647	719,108	8,955,065	2,229,368	20,210,171	Median
35,504,369	20,507,207	4,535,301	57,609,065	16,656,008	102,895,429	High
1,807,771	888,532	44,135	4,413,334	482,730	9,436,048	Low
610,305,773	505,065,985	98,420,446	1,213,792,203	340,762,295	2,683,008,943	Totals
112	112	112	112	112	113	Number of Libraries Responding
						NONUNIVERSITY LIBRARIES:
12,143,058	3,712,746	319,746	17,568,046	7,965,868	34,924,635	Median
383,658,502	37,578,811	2,913,139	473,003,880	378,124,103	913,539,942	Totals
7	6	6	10	10	10	Number of Libraries Responding
993,964,275	542,644,796	101,333,585	1,686,796,083	718,886,398	3,596,548,884	**GRAND TOTALS**

ARL LIBRARY DATA TABLES, 2004-05
EXPENDITURES FOR ELECTRONIC RESOURCES

(Survey Question #) INSTITUTION	Notes	Computer Files (21)	Electronic Serials (22)	Total E-Materials (21 + 22)	Total Materials (15)	E-Materials As % Of Lib. Materials
ALABAMA	BGLM+	194,920	2,442,789	2,637,709	6,274,306	42.04
ALBERTA	GLM+	775,930	5,548,425	6,324,354	12,689,026	49.84
ARIZONA	BGLM+	417,413	4,685,387	5,102,800	11,289,602	45.20
ARIZONA STATE	GL+	272,277	4,276,839	4,549,116	9,875,520	46.06
AUBURN	G+	235,500	1,961,692	2,197,192	5,261,396	41.76
BOSTON	GLM+	136,043	1,276,613	1,412,656	7,927,966	17.82
BOSTON COLLEGE	BGL+	1,085	2,109,367	2,110,452	7,783,477	27.11
BRIGHAM YOUNG	BGL+	UA/NA	2,098,655	2,098,655	10,301,595	20.37
BRITISH COLUMBIA	GLM+	210,263	5,237,506	5,447,769	11,794,175	46.19
BROWN	GM+	445,846	3,119,379	3,565,225	8,064,841	44.21
CALIFORNIA, BERKELEY	L+	UA/NA	UA/NA	UA/NA	15,532,371	UA/NA
CALIFORNIA, DAVIS	GLM+	641,583	2,503,258	3,144,841	6,417,684	49.00
CALIFORNIA, IRVINE	BGM+	1,651,845	1,829,553	3,481,398	6,926,009	50.27
CALIFORNIA, LOS ANGELES	GLM+	194,514	2,794,320	2,988,834	11,892,506	25.13
CALIFORNIA, RIVERSIDE	BG+	4,317	2,017,224	2,021,541	5,444,783	37.13
CALIFORNIA, SAN DIEGO	GM+	142,932	2,471,826	2,614,758	8,324,922	31.41
CALIFORNIA, SANTA BARBARA	+	10,831	1,791,420	1,802,251	5,712,246	31.55
CASE WESTERN RESERVE	GLM+	0	2,590,935	2,590,935	6,784,251	38.19
CHICAGO	BGLM+	849,688	2,906,124	3,755,812	13,489,603	27.84
CINCINNATI	GLM+	597,851	4,176,215	4,774,066	10,074,470	47.39
COLORADO	BG+	221,861	4,197,172	4,419,033	8,801,962	50.21
COLORADO STATE	G+	61,799	2,972,899	3,034,698	6,237,544	48.65
COLUMBIA	LM+	1,043,510	5,834,145	6,877,655	19,159,897	35.90
CONNECTICUT	GLM+	244,761	3,589,557	3,834,318	8,200,363	46.76
CORNELL	GL+	80,593	4,935,821	5,016,414	13,774,489	36.42
DARTMOUTH	GM+	229,818	3,690,444	3,920,262	7,238,345	54.16
DELAWARE	G+	65,861	4,442,201	4,508,062	7,364,975	61.21
DUKE	BLM+	296,277	3,441,055	3,737,332	12,279,569	30.44
EMORY	LM+	619,814	2,970,853	3,590,667	13,370,261	26.86
FLORIDA	BGLM+	69,844	4,328,324	4,398,168	10,698,308	41.11
FLORIDA STATE	BLM+	44,612	2,145,824	2,190,436	6,902,381	31.73
GEORGE WASHINGTON	BGLM+	306,298	1,900,251	2,206,549	9,172,967	24.05
GEORGETOWN	BGLM	134,440	2,103,139	2,237,579	9,371,271	23.88
GEORGIA	GL+	572,109	4,348,419	4,920,528	11,013,996	44.68
GEORGIA TECH	G+	UA/NA	UA/NA	UA/NA	5,294,007	UA/NA
GUELPH	G+	344,774	2,618,138	2,962,912	4,721,391	62.76
HARVARD	GLM+	4,370,397	UA/NA	4,370,397	27,569,823	15.85
HAWAII	G+	UA/NA	2,264,524	2,264,524	5,910,365	38.31
HOUSTON	BGL+	196,995	1,329,260	1,526,255	7,681,826	19.87
HOWARD	BLM	0	1,553,660	1,553,660	3,885,992	39.98
ILLINOIS, CHICAGO	GM+	371,730	2,546,740	2,918,470	7,729,836	37.76
ILLINOIS, URBANA	GL+	29,282	3,248,104	3,277,386	13,053,827	25.11

UA/NA – Unavailable or Not Applicable
L - Includes Law library
M- Includes Medical library

+ - See Footnotes
B - Basis of volume count is bibliographic
G - Government documents not included in serials count

ARL LIBRARY DATA TABLES, 2004-05
EXPENDITURES FOR ELECTRONIC RESOURCES

Bibl. Utilities, Networks, and Consortia		Hardware	Document Delivery/	
Library	External	And Software	Interlibrary Loan	
(23.a)	(23.b)	(24)	(25)	(Survey Question #) INSTITUTION
227,325	UA/NA	459,737	16,743	ALABAMA
UA/NA	UA/NA	UA/NA	UA/NA	ALBERTA
7,250	UA/NA	656,110	172,945	ARIZONA
477,925	0	900,460	193,189	ARIZONA STATE
148,963	150,000	517,807	258,186	AUBURN
135,299	43,536	258,330	66,632	BOSTON
146,500	UA/NA	144,712	4,432	BOSTON COLLEGE
455,552	UA/NA	693,448	45,000	BRIGHAM YOUNG
35,502	0	UA/NA	169,797	BRITISH COLUMBIA
220,876	0	305,835	179,810	BROWN
UA/NA	UA/NA	UA/NA	UA/NA	CALIFORNIA, BERKELEY
134,749	880,340	464,085	224,954	CALIFORNIA, DAVIS
135,828	2,034,775	390,173	71,417	CALIFORNIA, IRVINE
511,432	UA/NA	2,124,182	483,771	CALIFORNIA, LOS ANGELES
149,000	0	448,520	29,091	CALIFORNIA, RIVERSIDE
UA/NA	739,990	1,061,465	178,371	CALIFORNIA, SAN DIEGO
324,000	UA/NA	672,424	73,829	CALIFORNIA, SANTA BARBARA
119,983	139,554	504,083	29,328	CASE WESTERN RESERVE
251,395	250,000	471,663	189,619	CHICAGO
198,359	479,374	570,700	28,357	CINCINNATI
330,703	33,042	140,687	136,992	COLORADO
247,090	202,144	708,462	290,273	COLORADO STATE
634,032	UA/NA	1,049,369	92,468	COLUMBIA
328,761	1,894,324	502,641	173,312	CONNECTICUT
322,615	UA/NA	1,497,797	162,455	CORNELL
160,133	0	182,676	44,035	DARTMOUTH
86,704	0	474,289	56,336	DELAWARE
115,880	4,777	88,277	7,400	DUKE
293,343	219,299	537,299	62,634	EMORY
UA/NA	UA/NA	354,836	40,921	FLORIDA
74,576	1,498,925	347,221	51,308	FLORIDA STATE
740,388	90,485	457,951	82,109	GEORGE WASHINGTON
195,243	0	495,640	68,419	GEORGETOWN
422,551	0	420,322	37,388	GEORGIA
75,000	0	489,199	157,325	GEORGIA TECH
27,450	0	435,415	121,186	GUELPH
2,024,689	UA/NA	1,675,592	59,657	HARVARD
80,591	0	58,156	51,300	HAWAII
313,550	UA/NA	714,914	96,265	HOUSTON
186,604	UA/NA	33,664	40,104	HOWARD
11,928	UA/NA	UA/NA	23,083	ILLINOIS, CHICAGO
259,282	UA/NA	317,693	45,906	ILLINOIS, URBANA

UA/NA – Unavailable or Not Applicable

ARL Library Data Tables, 2004-05
Expenditures for Electronic Resources

(Survey Question #) INSTITUTION	Notes	Computer Files (21)	Electronic Serials (22)	Total E-Materials (21 + 22)	Total Materials (15)	E-Materials As % Of Lib. Materials
INDIANA	L+	1,192,071	2,929,367	4,121,438	12,674,170	32.52
IOWA	GLM+	199,402	3,844,783	4,044,185	11,866,373	34.08
IOWA STATE	G+	50,626	3,648,594	3,699,220	8,599,786	43.02
JOHNS HOPKINS	GM+	787,065	4,078,450	4,865,515	12,279,738	39.62
KANSAS	LM+	106,800	1,761,084	1,867,884	8,333,205	22.41
KENT STATE	G+	UA/NA	1,553,020	1,553,020	3,432,288	45.25
KENTUCKY	GLM+	235,566	2,490,096	2,725,662	9,413,427	28.96
LAVAL	GLM+	UA/NA	UA/NA	UA/NA	7,803,045	UA/NA
LOUISIANA STATE	LM+	88,985	968,175	1,057,160	6,932,439	15.25
LOUISVILLE	GLM+	0	2,171,175	2,171,175	8,641,096	25.13
MCGILL	GLM+	662,132	5,182,313	5,844,445	11,376,632	51.37
MCMASTER	GM+	123,879	3,007,021	3,130,900	5,984,915	52.31
MANITOBA	GLM+	415,896	2,462,307	2,878,204	6,363,752	45.23
MARYLAND	G+	98,070	4,049,376	4,147,446	8,415,165	49.29
MASSACHUSETTS	G+	0	1,954,626	1,954,626	5,278,206	37.03
MIT	G+	122,217	2,400,589	2,522,806	7,140,052	35.33
MIAMI	GLM+	210,889	2,621,286	2,832,175	10,643,708	26.61
MICHIGAN	LM+	UA/NA	UA/NA	UA/NA	20,309,007	UA/NA
MICHIGAN STATE	GL+	482,200	3,201,454	3,683,654	9,487,402	38.83
MINNESOTA	BGLM+	31,019	4,478,091	4,509,110	12,559,689	35.90
MISSOURI	BGLM+	185,801	2,073,906	2,259,707	5,833,439	38.74
MONTREAL	GLM+	235,628	3,798,418	4,034,046	9,517,765	42.38
NEBRASKA	GL+	725,904	1,617,186	2,343,090	6,572,861	35.65
NEW MEXICO	LM+	202,424	1,801,761	2,004,185	5,817,037	34.45
NEW YORK	GLM+	1,200,169	4,207,382	5,407,551	14,454,550	37.41
NORTH CAROLINA	GLM+	306,160	1,410,269	1,716,429	12,766,914	13.44
NORTH CAROLINA STATE	G+	523,762	1,683,886	2,207,648	8,873,695	24.88
NORTHWESTERN	GLM+	619,086	4,931,465	5,550,551	11,536,325	48.11
NOTRE DAME	GL+	847,382	3,167,243	4,014,625	8,662,668	46.34
OHIO	BM+	37,557	2,225,848	2,263,405	5,391,032	41.98
OHIO STATE	GLM+	76,694	4,500,542	4,577,236	8,880,060	51.55
OKLAHOMA	GLM+	874,222	1,606,358	2,480,580	11,268,898	22.01
OKLAHOMA STATE	BGM+	280,417	3,314,575	3,594,992	6,690,031	53.74
OREGON	BGL+	21,700	1,685,586	1,707,286	5,334,251	32.01
PENNSYLVANIA	LM+	28,443	4,060,345	4,088,788	13,100,485	31.21
PENNSYLVANIA STATE	GLM+	1,605,790	6,407,965	8,013,755	17,587,527	45.56
PITTSBURGH	GLM+	221,623	4,683,629	4,905,252	13,070,488	37.53
PRINCETON	G+	168,726	3,202,652	3,371,378	14,230,665	23.69
PURDUE	G+	UA/NA	2,252,541	2,252,541	9,542,018	23.61
QUEEN'S	BLM+	UA/NA	3,464,480	3,464,480	6,767,889	51.19
RICE	G+	813,948	1,796,580	2,610,528	8,878,432	29.40
ROCHESTER	BGM+	32,865	1,324,612	1,357,477	6,421,998	21.14

UA/NA – Unavailable or Not Applicable
L - Includes Law library
M- Includes Medical library

+ - See Footnotes
B - Basis of volume count is bibliographic
G - Government documents not included in serials count

EXPENDITURES FOR ELECTRONIC RESOURCES

Bibl. Utilities, Networks, and Consortia		Hardware	Document Delivery/	
Library	External	And Software	Interlibrary Loan	
(23.a)	(23.b)	(24)	(25)	(Survey Question #) INSTITUTION
186,457	229,540	672,136	102,809	INDIANA
266,441	0	271,575	233,055	IOWA
85,789	0	543,677	117,681	IOWA STATE
337,132	0	403,749	184,132	JOHNS HOPKINS
225,115	0	874,635	264,498	KANSAS
134,040	UA/NA	548,980	17,789	KENT STATE
372,837	0	638,263	204,359	KENTUCKY
UA/NA	UA/NA	UA/NA	UA/NA	LAVAL
348,850	108,096	92,315	272,620	LOUISIANA STATE
169,330	0	214,911	25,690	LOUISVILLE
182,035	0	811,696	198,923	MCGILL
66,343	0	201,975	80,904	MCMASTER
UA/NA	UA/NA	265,049	186,338	MANITOBA
319,935	148,791	22,745,914	9,681	MARYLAND
140,567	75,000	169,547	99,632	MASSACHUSETTS
126,269	0	406,490	130,641	MIT
261,247	0	587,680	131,073	MIAMI
UA/NA	UA/NA	UA/NA	UA/NA	MICHIGAN
222,205	UA/NA	753,296	236,011	MICHIGAN STATE
247,684	228,683	1,802,580	320,460	MINNESOTA
134,908	143,546	223,028	104,178	MISSOURI
113,085	0	688,700	119,556	MONTREAL
76,483	314,188	387,395	66,809	NEBRASKA
327,568	55,627	354,352	155,397	NEW MEXICO
96,080	3,150	851,475	178,524	NEW YORK
165,587	UA/NA	857,920	128,143	NORTH CAROLINA
168,823	0	632,222	139,934	NORTH CAROLINA STATE
238,910	UA/NA	667,381	89,899	NORTHWESTERN
260,331	151,247	472,455	107,067	NOTRE DAME
106,895	517,378	706,997	4,174	OHIO
256,403	1,170,543	552,732	110,037	OHIO STATE
222,577	59,124	269,762	72,657	OKLAHOMA
153,831	114,924	263,504	60,536	OKLAHOMA STATE
219,322	32,697	456,791	85,144	OREGON
652,955	0	1,289,500	190,189	PENNSYLVANIA
301,658	0	2,264,349	79,807	PENNSYLVANIA STATE
308,160	UA/NA	745,855	93,379	PITTSBURGH
349,600	0	1,370,979	207,079	PRINCETON
117,521	153,000	418,525	308,978	PURDUE
198,966	UA/NA	235,932	36,219	QUEEN'S
116,445	0	267,142	33,446	RICE
158,866	0	765,450	66,374	ROCHESTER

UA/NA - Unavailable or Not Applicable

ARL Library Data Tables, 2004-05
Expenditures for Electronic Resources

(Survey Question #) INSTITUTION	Notes	Computer Files (21)	Electronic Serials (22)	Total E-Materials (21 + 22)	Total Materials (15)	E-Materials As % Of Lib. Materials
RUTGERS	GL+	6,015	3,676,142	3,682,157	10,027,436	36.72
SASKATCHEWAN	GLM+	8,854	2,588,254	2,597,108	6,382,518	40.69
SOUTH CAROLINA	BLM+	55,842	2,855,603	2,911,445	6,364,082	45.75
SOUTHERN CALIFORNIA	GLM+	207,824	3,567,032	3,774,856	11,857,390	31.84
SOUTHERN ILLINOIS	LM+	218,902	2,416,752	2,635,654	7,073,858	37.26
SUNY-ALBANY	G+	10,075	1,523,912	1,533,987	5,122,321	29.95
SUNY-BUFFALO	GLM+	70,844	3,629,118	3,699,962	7,391,091	50.06
SUNY-STONY BROOK	BM+	13,873	2,328,017	2,341,890	6,217,224	37.67
SYRACUSE	GL+	191,485	2,403,859	2,595,344	5,351,232	48.50
TEMPLE	BGLM+	527,993	3,199,574	3,727,567	7,295,589	51.09
TENNESSEE	GLM+	113,548	2,099,885	2,213,433	10,146,013	21.82
TEXAS	GL+	615,641	3,884,365	4,500,006	13,149,394	34.22
TEXAS A&M	G+	516,108	6,283,975	6,800,083	14,469,892	46.99
TEXAS TECH	LM+	484,585	1,699,367	2,183,952	9,269,276	23.56
TORONTO	GL+	853,889	2,299,916	3,153,805	19,816,389	15.92
TULANE	LM+	138,179	1,807,362	1,945,541	6,525,782	29.81
UTAH	BGLM+	29,106	697,531	726,637	7,481,035	9.71
VANDERBILT	GLM+	172,362	3,955,139	4,127,501	9,967,495	41.41
VIRGINIA	LM+	508,430	2,684,658	3,193,088	10,082,602	31.67
VIRGINIA TECH	G+	2,042	2,670,825	2,672,867	5,927,765	45.09
WASHINGTON	BGLM+	473,591	3,808,502	4,282,093	12,432,102	34.44
WASHINGTON STATE	G+	88,475	2,756,350	2,844,825	5,761,048	49.38
WASHINGTON U.-ST. LOUIS	GLM+	219,058	4,096,671	4,315,729	9,969,715	43.29
WATERLOO	G+	977,416	3,108,773	4,086,189	6,194,668	65.96
WAYNE STATE	G+	118,521	5,775,845	5,894,366	7,665,646	76.89
WESTERN ONTARIO	GL+	571,968	5,144,141	5,716,109	8,869,507	64.45
WISCONSIN	GLM+	391,921	2,133,765	2,525,686	10,938,665	23.09
YALE	GLM+	1,119,958	3,744,850	4,864,808	31,194,924	15.59
YORK	GL+	134,839	3,008,752	3,143,591	7,586,224	41.44
BOSTON PUBLIC	G+	0	631,100	631,100	5,943,731	10.62
CISTI	G+	UA/NA	1,529,789	1,529,789	9,751,507	15.69
CTR FOR RESEARCH LIBS	+	2,345	UA/NA	2,345	835,742	0.28
LIBRARY OF CONGRESS	BG+	168,345	1,257,499	1,425,844	12,505,812	11.40
NATL AGRICULTURAL LIB	G+	0	2,614,758	2,614,758	3,540,298	73.86
LIB & ARCHIVES CANADA	G+	47,513	27,611	75,124	1,400,708	5.36
NATL LIB OF MEDICINE	G+	5,167	159,089	164,256	8,264,442	1.99
NEW YORK PUBLIC	G+	0	1,182,795	1,182,795	11,871,901	9.96
NEW YORK STATE	+	UA/NA	UA/NA	UA/NA	3,322,506	UA/NA
SMITHSONIAN	+	0	124,689	124,689	2,770,731	4.50

UA/NA – Unavailable or Not Applicable
L - Includes Law library
M- Includes Medical library
+ - See Footnotes
B - Basis of volume count is bibliographic
G - Government documents not included in serials count

ARL Library Data Tables, 2004-05
Expenditures for Electronic Resources

Bibl. Utilities, Networks, and Consortia		Hardware	Document Delivery/	
Library	External	And Software	Interlibrary Loan	
(23.a)	(23.b)	(24)	(25)	(Survey Question #) INSTITUTION
323,342	UA/NA	259,638	68,981	RUTGERS
52,342	UA/NA	338,247	100,517	SASKATCHEWAN
107,877	0	370,689	110,002	SOUTH CAROLINA
1,020,803	0	2,116,322	114,641	SOUTHERN CALIFORNIA
169,164	62,466	306,330	90,891	SOUTHERN ILLINOIS
0	44,676	173,867	95,896	SUNY-ALBANY
UA/NA	UA/NA	UA/NA	UA/NA	SUNY-BUFFALO
675,896	332,353	371,337	54,955	SUNY-STONY BROOK
23,000	UA/NA	247,651	126,430	SYRACUSE
146,921	9,000	447,602	36,614	TEMPLE
74,841	37,978	544,871	174,708	TENNESSEE
157,738	88	1,454,211	208,691	TEXAS
8,508	399,583	884,630	387,351	TEXAS A&M
259,652	54,750	1,021,123	81,283	TEXAS TECH
341,429	UA/NA	3,304,111	150,426	TORONTO
152,405	24,239	197,227	15,015	TULANE
247,412	1,418,992	1,299,362	53,631	UTAH
304,643	0	510,699	151,905	VANDERBILT
182,856	685,895	1,786,410	148,008	VIRGINIA
93,822	0	598,386	226,455	VIRGINIA TECH
314,839	144,303	113,836	283,000	WASHINGTON
225,343	207,520	231,571	UA/NA	WASHINGTON STATE
242,392	9,694	2,116,443	83,455	WASHINGTON U.-ST. LOUIS
0	0	173,639	69,359	WATERLOO
78,443	0	759,173	76,556	WAYNE STATE
85,804	0	278,105	UA/NA	WESTERN ONTARIO
299,674	485,154	2,272,817	327,043	WISCONSIN
509,885	UA/NA	993,017	139,894	YALE
55,834	UA/NA	246,138	5,635	YORK
347,994	147,869	510,498	5,771	BOSTON PUBLIC
UA/NA	UA/NA	2,632,753	UA/NA	CISTI
263,967	0	67,553	143,945	CTR FOR RESEARCH LIBS
959,805	0	16,054,052	0	LIBRARY OF CONGRESS
0	771,573	760,273	1,284,964	NATL AGRICULTURAL LIB
34,310	0	0	24,748	LIB & ARCHIVES CANADA
UA/NA	UA/NA	388,536	2,269,052	NATL LIB OF MEDICINE
848,573	UA/NA	660,565	973,131	NEW YORK PUBLIC
1,800	20,875	UA/NA	UA/NA	NEW YORK STATE
55,484	0	0	4,464	SMITHSONIAN

UA/NA - Unavailable or Not Applicable

ARL Library Data Tables, 2004-05
Summary Data: Expenditures for Electronic Resources

	Computer Files	Electronic Serials	Total E-Materials	Total Materials	E-Materials As % Of Lib. Materials
(Survey Question #)	(21)	(22)	(21 + 22)	(15)	
UNIVERSITY LIBRARIES:					
Median	210,576	2,824,962	3,144,841	8,662,668	37.53
High	4,370,397	6,407,965	8,013,755	31,194,924	76.89
Low	0	697,531	726,637	3,432,288	9.71
Totals	38,744,076	328,166,027	366,910,103	1,080,558,152	
Number of Libraries Responding	104	108	109	113	109
NONUNIVERSITY LIBRARIES:					
Median	1,173	906,948	631,100	4,742,015	9.96
Totals	223,370	7,527,331	7,750,700	60,207,378	
Number of Libraries Responding	8	8	9	10	9
GRAND TOTALS	38,967,446	335,693,358	374,660,803	1,140,765,530	

ARL Library Data Tables, 2004-05
Summary Data: Expenditures for Electronic Resources

Bibl. Utilities, Networks, and Consortia: Library	Bibl. Utilities, Networks, and Consortia: External	Hardware And Software	Document Delivery/ Interlibrary Loan	
(23.a)	(23.b)	(24)	(25)	(Survey Question #)
				UNIVERSITY LIBRARIES:
186,604	24,239	492,420	100,074	Median
2,024,689	2,034,775	22,745,914	483,771	High
0	0	33,664	4,174	Low
25,203,164	16,082,790	91,790,199	12,951,510	Totals
105	81	106	106	Number of Libraries Responding
				NONUNIVERSITY LIBRARIES:
159,726	0	510,498	84,347	Median
2,511,933	940,317	21,074,230	4,706,075	Totals
8	7	9	8	Number of Libraries Responding
27,715,097	17,023,107	112,864,430	17,657,585	**GRAND TOTALS**

ARL Library Data Tables, 2004-05
Personnel and Public Services

(Survey Question #) INSTITUTION	Notes	Professional Staff (FTE) (26.a)	Support Staff (FTE) (26.b)	Student Assistants (FTE) (26.c)	Total Staff (FTE) (26)	Staffed Service Points (27)	Service Hours Per Week (28)
ALABAMA	BGLM+	60	81	65	206	20	110
ALBERTA	GLM+	80	214	23	317	23	101
ARIZONA	BGLM+	100	156	84	340	21	168
ARIZONA STATE	GL+	90	207	43	340	22	149
AUBURN	G+	51	57	36	144	11	110
BOSTON	GLM+	95	102	110	307	33	108
BOSTON COLLEGE	BGL+	76	77	53	206	14	110
BRIGHAM YOUNG	BGL+	114	57	209	380	21	101
BRITISH COLUMBIA	GLM+	99	199	43	341	32	101
BROWN	GM+	80	101	32	213	11	149
CALIFORNIA, BERKELEY	L+	212	188	224	624	UA/NA	UA/NA
CALIFORNIA, DAVIS	GLM+	62	150	57	269	15	95
CALIFORNIA, IRVINE	BGM+	61	121	60	242	16	120
CALIFORNIA, LOS ANGELES	GLM+	169	282	153	604	32	95
CALIFORNIA, RIVERSIDE	BG+	43	90	30	163	14	91
CALIFORNIA, SAN DIEGO	GM+	92	185	74	351	18	168
CALIFORNIA, SANTA BARBARA	+	49	137	40	226	10	103
CASE WESTERN RESERVE	GLM+	63	55	32	150	16	109
CHICAGO	BGLM+	64	178	66	308	32	144
CINCINNATI	GLM+	111	75	54	240	25	101
COLORADO	BG+	51	117	49	217	24	104
COLORADO STATE	G+	46	60	31	137	6	108
COLUMBIA	LM+	239	263	100	602	50	106
CONNECTICUT	GLM+	79	82	59	220	23	114
CORNELL	GL+	119	313	108	540	51	144
DARTMOUTH	GM+	58	116	28	202	20	115
DELAWARE	G+	57	107	39	203	16	100
DUKE	BLM+	156	154	32	342	22	119
EMORY	LM+	93	159	47	299	25	113
FLORIDA	BGLM+	116	196	123	435	20	104
FLORIDA STATE	BLM+	68	129	48	245	22	118
GEORGE WASHINGTON	BGLM+	74	120	55	249	26	111
GEORGETOWN	BGLM	83	132	50	265	20	153
GEORGIA	GL+	79	185	44	308	20	110
GEORGIA TECH	G+	48	70	7	125	8	135
GUELPH	G+	39	74	18	131	14	133
HARVARD	GLM+	504	583	111	1,198	UA/NA	UA/NA
HAWAII	G+	69	89	44	202	14	98
HOUSTON	BGL+	87	68	43	198	8	108
HOWARD	BLM	54	75	38	167	18	144
ILLINOIS, CHICAGO	GM+	72	123	83	278	22	93
ILLINOIS, URBANA	GL+	186	202	127	515	50	119

UA/NA – Unavailable or Not Applicable
L - Includes Law library
M- Includes Medical library
+ - See Footnotes
B - Basis of volume count is bibliographic
G - Government documents not included in serials count
* Figures in italics are derived from a sampling method rather than an actual annual count.

ARL Library Data Tables, 2004-05
Personnel and Public Services

Presentations Groups*	Presentations Participants*	Reference Transactions*	Initial Circulation	Total Circulation	ILL/DD Loaned	ILL/DD Borrowed	(Survey Question #) INSTITUTION
(29)	(30)	(31)	(32)	(33)	(34)	(35)	
562	10,754	*92,605*	177,912	229,577	12,053	9,997	ALABAMA
938	21,898	*124,027*	UA/NA	645,277	97,042	33,776	ALBERTA
2,117	37,163	64,860	413,771	655,486	50,161	29,212	ARIZONA
740	20,203	141,953	410,610	1,155,724	59,341	27,541	ARIZONA STATE
690	16,571	*103,844*	178,172	244,036	12,456	8,748	AUBURN
854	11,267	84,714	290,512	398,718	17,254	18,007	BOSTON
753	10,981	33,848	181,560	331,052	19,973	12,473	BOSTON COLLEGE
2,833	35,668	189,906	619,358	784,811	34,391	38,362	BRIGHAM YOUNG
1,280	24,068	133,077	797,264	3,856,268	24,718	23,666	BRITISH COLUMBIA
366	UA/NA	*33,210*	212,362	298,982	36,325	34,474	BROWN
1,547	23,115	111,637	764,882	UA/NA	46,310	22,800	CALIFORNIA, BERKELEY
807	12,323	84,293	301,983	379,984	33,112	35,948	CALIFORNIA, DAVIS
817	18,587	*59,604*	217,409	439,969	29,067	28,323	CALIFORNIA, IRVINE
2,904	21,170	86,075	630,888	2,511,234	59,765	41,077	CALIFORNIA, LOS ANGELES
501	9,025	*31,082*	187,314	256,341	23,429	16,827	CALIFORNIA, RIVERSIDE
1,802	12,719	93,395	468,973	694,991	56,644	28,753	CALIFORNIA, SAN DIEGO
572	9,940	*102,544*	256,252	428,551	24,637	32,804	CALIFORNIA, SANTA BARBARA
561	6,196	22,205	111,414	144,108	40,160	39,709	CASE WESTERN RESERVE
699	9,242	*43,755*	496,211	758,784	46,516	24,920	CHICAGO
2,018	23,344	138,490	289,824	422,222	57,635	50,115	CINCINNATI
890	14,624	*158,712*	434,459	691,499	51,802	29,054	COLORADO
411	7,863	*103,752*	233,810	420,495	49,178	65,964	COLORADO STATE
2,120	25,464	*224,947*	506,972	1,134,681	54,523	53,142	COLUMBIA
1,407	14,604	61,702	190,207	219,668	53,812	34,709	CONNECTICUT
1,451	22,149	*102,820*	535,241	1,027,104	59,431	32,339	CORNELL
UA/NA	UA/NA	UA/NA	235,189	310,093	34,139	29,766	DARTMOUTH
453	9,460	*131,701*	UA/NA	351,213	17,323	20,933	DELAWARE
1,018	11,706	*126,921*	390,086	534,244	41,107	21,044	DUKE
785	12,273	55,457	251,741	539,963	42,853	17,634	EMORY
717	14,856	*194,622*	566,604	594,747	21,411	24,955	FLORIDA
605	11,503	67,432	303,588	420,205	45,473	20,733	FLORIDA STATE
1,023	13,851	*51,682*	210,856	237,988	36,864	58,194	GEORGE WASHINGTON
1,042	16,448	*95,893*	266,766	387,995	37,201	22,604	GEORGETOWN
816	15,031	*115,410*	318,884	482,117	51,995	9,378	GEORGIA
150	3,502	12,553	93,799	UA/NA	9,269	5,045	GEORGIA TECH
422	6,685	22,376	196,608	436,825	46,319	66,594	GUELPH
UA/NA	UA/NA	UA/NA	UA/NA	2,366,950	44,721	14,880	HARVARD
96	1,632	63,551	283,442	457,976	2,513	8,178	HAWAII
412	11,148	*62,272*	271,090	310,152	29,435	19,463	HOUSTON
402	*5,787*	*160,000*	51,592	75,719	16,069	5,102	HOWARD
654	9,112	59,063	154,038	366,005	74,438	39,368	ILLINOIS, CHICAGO
1,090	15,650	*234,642*	502,438	646,908	90,304	91,596	ILLINOIS, URBANA

UA/NA - Unavailable or Not Applicable

* Figures in italics are derived from a sampling method rather than an actual annual count.

ARL Library Data Tables, 2004-05
Personnel and Public Services

(Survey Question #) INSTITUTION	Notes	Professional Staff (FTE) (26.a)	Support Staff (FTE) (26.b)	Student Assistants (FTE) (26.c)	Total Staff (FTE) (26)	Staffed Service Points (27)	Service Hours Per Week (28)
INDIANA	L+	163	180	99	442	64	168
IOWA	GLM+	89	140	69	298	36	124
IOWA STATE	G+	51	93	40	184	13	105
JOHNS HOPKINS	GM+	135	161	41	337	31	168
KANSAS	LM+	103	109	80	292	11	142
KENT STATE	G+	32	48	40	120	20	102
KENTUCKY	GLM+	97	119	65	281	23	140
LAVAL	GLM+	60	155	7	222	17	88
LOUISIANA STATE	LM+	51	99	45	195	28	93
LOUISVILLE	GLM+	51	91	45	187	21	103
MCGILL	GLM+	66	163	47	276	33	86
MCMASTER	GM+	34	106	23	163	12	100
MANITOBA	GLM+	61	139	27	227	31	100
MARYLAND	G+	127	90	50	267	8	129
MASSACHUSETTS	G+	55	75	46	176	14	106
MIT	G+	96	95	15	206	11	104
MIAMI	GLM+	92	115	66	273	20	117
MICHIGAN	LM+	160	313	123	596	UA/NA	UA/NA
MICHIGAN STATE	GL+	74	121	95	290	23	148
MINNESOTA	BGLM+	102	214	88	404	54	104
MISSOURI	BGLM+	56	115	29	200	21	97
MONTREAL	GLM+	110	282	2	394	65	101
NEBRASKA	GL+	52	114	45	211	20	96
NEW MEXICO	LM+	69	160	54	283	18	103
NEW YORK	GLM+	119	226	88	433	32	119
NORTH CAROLINA	GLM+	141	205	84	430	29	145
NORTH CAROLINA STATE	G+	124	108	56	288	17	146
NORTHWESTERN	GLM+	116	139	85	340	29	119
NOTRE DAME	GL+	58	158	36	252	18	126
OHIO	BM+	64	68	51	183	18	146
OHIO STATE	GLM+	134	156	133	423	43	168
OKLAHOMA	GLM+	61	92	67	220	25	117
OKLAHOMA STATE	BGM+	63	80	70	213	20	119
OREGON	BGL+	53	99	63	215	22	96
PENNSYLVANIA	LM+	119	177	104	400	32	100
PENNSYLVANIA STATE	GLM+	149	399	77	625	61	168
PITTSBURGH	GLM+	118	167	72	357	57	118
PRINCETON	G+	145	210	43	398	37	115
PURDUE	G+	66	133	56	255	23	119
QUEEN'S	BLM+	45	110	17	172	14	115
RICE	G+	57	63	14	134	12	143
ROCHESTER	BGM+	92	67	41	200	21	119

UA/NA – Unavailable or Not Applicable
+ - See Footnotes
L - Includes Law library
B - Basis of volume count is bibliographic
M- Includes Medical library
G - Government documents not included in serials count
* Figures in italics are derived from a sampling method rather than an actual annual count.

ARL LIBRARY DATA TABLES, 2004-05
PERSONNEL AND PUBLIC SERVICES

Presentations Groups*	Presentations Participants*	Reference Transactions*	Initial Circulation	Total Circulation	ILL/DD Loaned	ILL/DD Borrowed	(Survey Question #) INSTITUTION
(29)	(30)	(31)	(32)	(33)	(34)	(35)	
1,086	28,081	*275,360*	759,029	1,382,838	86,289	41,826	INDIANA
724	16,499	*99,216*	UA/NA	610,445	58,129	27,744	IOWA
432	9,406	31,665	256,692	294,873	34,603	16,389	IOWA STATE
644	8,425	85,157	279,224	483,005	37,960	36,374	JOHNS HOPKINS
1,155	19,436	*71,071*	426,227	555,330	51,672	38,009	KANSAS
830	15,286	*42,690*	284,643	377,963	12,825	8,420	KENT STATE
996	16,726	83,018	275,588	342,746	35,541	29,300	KENTUCKY
411	7,414	135,718	526,070	839,597	21,326	15,520	LAVAL
956	5,174	46,834	123,427	247,375	9,574	28,199	LOUISIANA STATE
709	11,460	64,669	UA/NA	385,220	24,666	12,119	LOUISVILLE
1,165	17,817	120,447	757,475	1,105,238	17,475	25,718	MCGILL
636	10,620	57,779	245,334	543,156	16,093	15,370	MCMASTER
971	9,698	84,599	296,154	438,509	27,540	38,329	MANITOBA
1,231	25,324	*353,526*	384,563	584,319	16,818	40,017	MARYLAND
296	5,146	58,333	264,421	408,867	41,537	34,507	MASSACHUSETTS
293	5,881	*42,326*	220,452	382,470	12,518	11,443	MIT
1,779	16,136	*86,593*	194,369	226,846	33,755	16,918	MIAMI
UA/NA	UA/NA	UA/NA	UA/NA	UA/NA	UA/NA	UA/NA	MICHIGAN
625	17,813	42,072	325,812	807,538	48,702	31,527	MICHIGAN STATE
1,129	18,481	168,098	424,555	602,585	180,504	35,615	MINNESOTA
763	*9,813*	*46,559*	247,280	315,620	58,619	44,953	MISSOURI
1,223	14,354	124,085	549,877	981,004	27,800	19,926	MONTREAL
236	6,383	49,736	193,822	444,309	28,123	15,260	NEBRASKA
1,000	16,555	*57,340*	284,699	377,163	21,450	22,114	NEW MEXICO
925	14,964	125,479	461,638	1,022,590	15,236	24,607	NEW YORK
1,318	22,296	129,162	UA/NA	1,053,054	54,840	14,947	NORTH CAROLINA
543	14,854	*59,662*	309,977	407,178	13,003	26,813	NORTH CAROLINA STATE
730	8,132	118,169	206,231	326,025	32,515	36,275	NORTHWESTERN
485	7,743	26,468	191,557	328,122	23,323	16,011	NOTRE DAME
946	18,381	95,864	365,774	518,162	88,448	71,084	OHIO
915	16,698	*315,600*	581,214	1,897,098	145,272	101,722	OHIO STATE
803	11,287	65,168	197,742	231,333	30,669	34,719	OKLAHOMA
1,040	32,641	167,113	432,949	482,973	37,506	20,089	OKLAHOMA STATE
833	11,619	*67,730*	266,442	335,551	75,939	61,671	OREGON
1,036	20,269	*227,638*	390,669	550,110	41,307	72,214	PENNSYLVANIA
2,893	46,587	194,722	631,139	668,877	57,373	53,146	PENNSYLVANIA STATE
1,872	25,934	*323,881*	372,003	499,939	74,175	22,394	PITTSBURGH
501	5,941	*37,991*	372,015	461,008	34,790	32,341	PRINCETON
326	7,776	53,274	182,776	596,803	29,018	27,862	PURDUE
674	10,658	29,986	246,411	502,474	26,157	13,349	QUEEN'S
105	1,242	UA/NA	127,971	213,794	12,704	12,984	RICE
764	10,000	*59,015*	196,797	239,762	26,482	21,231	ROCHESTER

UA/NA - Unavailable or Not Applicable
* Figures in italics are derived from a sampling method rather than an actual annual count.

ARL Library Data Tables, 2004-05
Personnel and Public Services

(Survey Question #) INSTITUTION	Notes	Professional Staff (FTE) (26.a)	Support Staff (FTE) (26.b)	Student Assistants (FTE) (26.c)	Total Staff (FTE) (26)	Staffed Service Points (27)	Service Hours Per Week (28)
RUTGERS	GL+	105	240	116	461	42	104
SASKATCHEWAN	GLM+	47	99	13	159	17	90
SOUTH CAROLINA	BLM+	68	108	94	270	17	104
SOUTHERN CALIFORNIA	GLM+	125	122	129	376	28	160
SOUTHERN ILLINOIS	LM+	56	93	62	211	15	112
SUNY-ALBANY	G+	59	70	41	170	12	109
SUNY-BUFFALO	GLM+	137	50	78	265	18	141
SUNY-STONY BROOK	BM+	90	29	42	161	11	100
SYRACUSE	GL+	65	135	27	227	22	104
TEMPLE	BGLM+	62	91	59	212	29	95
TENNESSEE	GLM+	92	141	53	286	30	168
TEXAS	GL+	149	311	112	572	29	107
TEXAS A&M	G+	132	141	93	366	18	125
TEXAS TECH	LM+	100	119	101	320	29	130
TORONTO	GL+	185	351	127	663	100	129
TULANE	LM+	54	106	32	192	18	116
UTAH	BGLM+	76	195	85	356	21	116
VANDERBILT	GLM+	103	108	41	252	21	143
VIRGINIA	LM+	102	214	58	374	31	149
VIRGINIA TECH	G+	36	90	35	161	9	102
WASHINGTON	BGLM+	138	214	116	468	42	135
WASHINGTON STATE	G+	47	84	36	167	20	91
WASHINGTON U.-ST. LOUIS	GLM+	108	133	57	298	23	116
WATERLOO	G+	35	99	43	177	15	106
WAYNE STATE	G+	99	89	77	265	19	142
WESTERN ONTARIO	GL+	73	102	24	199	9	107
WISCONSIN	GLM+	229	177	182	588	54	148
YALE	GLM+	254	370	78	702	44	111
YORK	GL+	60	112	61	233	20	93
BOSTON PUBLIC	G+	196	243	62	501	96	68
CISTI	G+	168	160	4	332	3	40
CTR FOR RESEARCH LIBS	+	25	14	26	65	2	40
LIBRARY OF CONGRESS	BG+	1,442	2,632	16	4,090	38	65
NATL AGRICULTURAL LIB	G+	131	82	2	215	13	45
LIB & ARCHIVES CANADA	G+	656	469	14	1,139	7	43
NATL LIB OF MEDICINE	G+	180	80	13	273	5	46
NEW YORK PUBLIC	G+	265	288	98	651	42	46
NEW YORK STATE	+	60	95	6	161	5	40
SMITHSONIAN	+	65	55	(1)	120	31	168

UA/NA – Unavailable or Not Applicable
L - Includes Law library
M- Includes Medical library
+ - See Footnotes
B - Basis of volume count is bibliographic
G - Government documents not included in serials count
* Figures in italics are derived from a sampling method rather than an actual annual count.

ARL Library Data Tables, 2004-05
Personnel and Public Services

Presentations: Groups* (29)	Presentations: Participants* (30)	Reference Transactions* (31)	Initial Circulation (32)	Total Circulation (33)	ILL/DD: Loaned (34)	ILL/DD: Borrowed (35)	(Survey Question #) INSTITUTION
1,113	22,395	115,816	317,090	651,060	25,699	25,729	RUTGERS
366	5,697	28,131	300,400	409,418	25,924	24,295	SASKATCHEWAN
553	8,039	160,899	981,922	1,020,093	21,503	21,166	SOUTH CAROLINA
1,056	11,711	60,005	388,308	614,261	24,066	11,351	SOUTHERN CALIFORNIA
1,103	8,369	70,204	249,651	464,314	53,688	32,660	SOUTHERN ILLINOIS
682	5,055	82,161	99,106	144,105	10,662	20,088	SUNY-ALBANY
702	16,077	62,084	347,442	393,075	37,536	20,440	SUNY-BUFFALO
612	7,893	63,711	181,101	238,375	23,160	14,749	SUNY-STONY BROOK
677	12,289	*64,083*	180,301	255,745	15,739	23,957	SYRACUSE
574	12,418	66,511	197,870	278,030	21,599	16,491	TEMPLE
929	14,921	87,195	253,911	335,332	37,938	30,075	TENNESSEE
1,452	61,042	165,755	1,620,495	2,640,064	51,621	30,097	TEXAS
498	14,291	29,909	739,946	1,374,651	45,758	61,658	TEXAS A&M
810	20,460	80,305	231,901	367,174	31,270	30,851	TEXAS TECH
2,121	30,328	578,220	1,468,103	2,043,679	36,580	12,127	TORONTO
445	6,203	22,793	160,856	271,772	24,484	12,701	TULANE
1,717	33,487	*355,798*	229,258	585,529	42,595	24,603	UTAH
692	10,257	37,801	212,710	348,541	27,692	19,547	VANDERBILT
1,334	14,905	*119,194*	634,236	1,068,288	42,612	24,344	VIRGINIA
607	13,484	23,057	243,554	290,854	21,585	21,218	VIRGINIA TECH
1,030	23,776	*150,924*	UA/NA	1,471,685	79,369	50,498	WASHINGTON
803	14,079	49,628	219,010	382,213	UA/NA	UA/NA	WASHINGTON STATE
862	*7,466*	*58,129*	278,705	514,913	40,603	51,110	WASHINGTON U.-ST. LOUIS
202	2,975	28,611	256,608	621,603	72,073	46,513	WATERLOO
749	16,597	55,583	141,819	228,326	58,149	29,966	WAYNE STATE
917	17,705	55,516	561,299	833,675	18,307	5,780	WESTERN ONTARIO
2,248	28,333	UA/NA	590,793	933,496	109,100	65,313	WISCONSIN
1,392	10,821	*96,329*	UA/NA	669,297	43,006	28,800	YALE
789	22,776	182,469	544,932	1,111,665	14,783	4,921	YORK
9,496	140,589	996,127	2,431,222	2,431,222	38,958	11,033	BOSTON PUBLIC
UA/NA	1,677	22,666	16,714	35,342	618,724	27,652	CISTI
47	*1,060*	UA/NA	UA/NA	71,632	24,467	UA/NA	CTR FOR RESEARCH LIBS
1,305	42,538	996,771	123,134	UA/NA	22,848	31	LIBRARY OF CONGRESS
434	45,766	14,453	5,001	6,145	19,921	9,682	NATL AGRICULTURAL LIB
1,578	*121,100*	*76,647*	193,315	205,642	79,957	1,076	LIB & ARCHIVES CANADA
274	6,684	*95,791*	238,833	UA/NA	341,239	205	NATL LIB OF MEDICINE
685	5,428	651,322	UA/NA	UA/NA	3,821	1,514	NEW YORK PUBLIC
168	1,634	79,577	34,207	39,989	22,007	1,014	NEW YORK STATE
236	3,281	71,143	33,759	234,627	7,031	4,509	SMITHSONIAN

UA/NA – Unavailable or Not Applicable
* Figures in italics are derived from a sampling method rather than an actual annual count.

ARL Library Data Tables, 2004-05
Summary Data: Personnel and Public Services

	Professional Staff (FTE)	Support Staff (FTE)	Student Assistants (FTE)	Total Staff (FTE)	Staffed Service Points	Service Hours Per Week
(Survey Question #)	(26.a)	(26.b)	(26.c)	(26)	(27)	(28)
UNIVERSITY LIBRARIES:						
Median	80	120	54	265	21	113
High	504	583	224	1,198	100	168
Low	32	29	2	120	6	86
Totals	10,744	16,276	7,242	34,262	2,732	
Number of Libraries Responding	113	113	113	113	110	110
NONUNIVERSITY LIBRARIES:						
Median	174	128	14	303	10	46
Totals	3,188	4,118	241	7,547	242	
Number of Libraries Responding	10	10	9	10	10	10
GRAND TOTALS	13,932	20,394	7,483	41,809	2,974	

ARL Library Data Tables, 2004-05
Summary Data: Personnel and Public Services

Presentations Groups*	Presentations Participants*	Reference Transactions*	Initial Circulation	Total Circulation	ILL/DD Loaned	ILL/DD Borrowed	
(29)	(30)	(31)	(32)	(33)	(34)	(35)	(Survey Question #)
							UNIVERSITY LIBRARIES:
805	13,851	82,590	281,333	459,492	35,541	25,729	Median
2,904	61,042	578,220	1,620,495	3,856,268	180,504	101,722	High
96	1,242	12,553	51,592	75,719	2,513	4,921	Low
102,359	1,650,381	11,155,266	37,488,426	70,281,764	4,470,546	3,278,311	Totals
110	109	108	104	110	111	111	Number of Libraries Responding
							NONUNIVERSITY LIBRARIES:
434	6,056	79,577	78,671	71,632	23,658	1,514	Median
14,223	369,757	3,004,497	3,076,185	3,024,599	1,178,973	56,716	Totals
9	10	9	8	7	10	9	Number of Libraries Responding
116,582	2,020,138	14,159,763	40,564,611	73,306,363	5,649,519	3,335,027	**GRAND TOTALS**

ANALYSIS OF SELECTED VARIABLES OF ARL UNIVERSITY LIBRARIES, 2004-05

The percentages and ratios below are select indicators that describe the condition of ARL university libraries. The high and low figures indicate the range, while the mean and median indicate the central tendency of the distributions for ARL university libraries. Note that a distribution is normal when the mean and the median figures are identical. If the mean is larger than the median then the distribution is positively skewed, and if the opposite is true the distribution is negatively skewed.

	Category	High	Mean	Median	Low	Number of libraries reporting
1.	Professional Staff as percent of Total Staff	55.90	31.38	29.56	19.77	113
2.	Support Staff as percent of Total Staff	71.57	47.31	47.42	15.00	113
3.	Student Assistant Staff as percent of Total Staff	55.00	21.31	21.72	0.51	113
4.	Ratio of Professional to Support Staff (excluding Student Assistant Staff)	3.10	0.74	0.64	0.32	113
5.	Ratio of Items Loaned to Items Borrowed	5.54	1.53	1.39	0.31	111
6.	Serial Expenditures as percent of Total Library Materials Expenditures	90.11	67.35	68.65	25.30	112
7.	Monograph Expenditures as percent of Total Library Materials Expenditures	47.55	22.52	21.05	5.87	112
8.	Total Library Materials Expenditures as percent of Total Library Expenditures	60.68	42.12	42.78	25.85	113
9.	Contract Binding as percent of Total Library Expenditures	1.94	0.76	0.70	0.22	112
10.	Salary and Wages Expenditures as percent of Total Library Expenditures	60.21	44.95	44.72	30.20	112
11.	Other Operating Expenditures as percent of Total Library Expenditures	41.44	12.19	11.48	2.95	112
12.	Unit cost of monographs (per volume)	136.25	56.81	55.64	24.17	100
13.	Unit cost of serials (per subscription)	656.31	247.55	231.90	93.72	88
14.	Library Expenditures per Faculty	48,195.75	16,941.82	14,660.38	7,227.49	112
15.	Library Expenditures per Full-Time Student	6,172.77	1,242.62	910.33	460.87	111
16.	Library Expenditures per Full-Time Graduate Student	18,361.53	5,311.16	4,816.34	2,006.16	112
17.	Library Expenditures per Full-Time User	5,340.79	1,144.09	860.77	440.82	111
18.	Library Expenditures per User (includes part-time)	5,253.10	981.82	692.90	361.85	111

PH.D., FACULTY, AND ENROLLMENT STATISTICS *

	Ph.D.s Awarded	Ph.D. Fields	Faculty	ENROLLMENTS Total FTE Full-time**	Total Part-time**	Graduate FTE Full-time	Graduate Part-time
(Survey Question #) INSTITUTION	(36)	(37)	(38)	(39)	(40)	(41)	(42)
ALABAMA	192	44	811	17,424	3,505	2,569	1,792
ALBERTA	272	116	1,512	30,605	4,012	4,064	1,816
ARIZONA	398	83	1,432	30,497	6,435	5,902	2,662
ARIZONA STATE	256	45	2,131	44,392	13,764	6,787	4,699
AUBURN	158	34	1,177	19,732	3,196	2,314	1,718
BOSTON	192	61	2,269	24,457	5,644	7,885	3,561
BOSTON COLLEGE	117	30	650	11,758	2,770	2,414	2,341
BRIGHAM YOUNG	68	22	1,287	24,814	6,033	1,360	1,448
BRITISH COLUMBIA	330	86	2,048	29,573	12,759	6,507	1,371
BROWN	151	38	628	7,604	400	1,896	94
CALIFORNIA, BERKELEY	874	82	1,429	29,379	2,532	7,437	1,268
CALIFORNIA, DAVIS	377	62	1,870	27,345	2,720	6,454	498
CALIFORNIA, IRVINE	211	40	1,350	23,109	1,235	3,965	517
CALIFORNIA, LOS ANGELES	665	80	2,460	36,268	1,295	12,617	354
CALIFORNIA, RIVERSIDE	159	40	1,742	16,530	574	1,980	35
CALIFORNIA, SAN DIEGO	304	64	1,675	22,690	1,053	3,306	98
CALIFORNIA, SANTA BARBARA	287	42	976	20,408	618	2,879	42
CASE WESTERN RESERVE	187	57	592	7,043	2,052	3,791	1,788
CHICAGO	327	64	1,295	11,392	2,478	6,879	2,446
CINCINNATI	188	56	1,943	20,860	6,318	5,002	3,048
COLORADO	256	43	1,115	26,202	6,160	2,037	3,895
COLORADO STATE	187	41	934	21,240	5,023	1,946	2,588
COLUMBIA	513	108	1,931	22,300	6,704	13,790	5,694
CONNECTICUT	261	67	1,541	21,664	5,615	4,130	2,998
CORNELL	406	82	1,675	19,471	47	5,874	19
DARTMOUTH	56	15	781	5,561	143	1,023	91
DELAWARE	189	43	1,070	18,134	3,104	2,519	876
DUKE	259	43	956	12,382	442	6,076	393
EMORY	160	28	742	11,287	847	4,866	758
FLORIDA	702	108	1,622	38,755	5,520	7,772	2,809
FLORIDA STATE	271	137	1,126	32,774	6,878	5,237	2,689
GEORGE WASHINGTON	95	28	1,031	15,567	8,525	5,770	7,355
GEORGETOWN	80	21	1,165	11,970	1,263	5,656	1,055
GEORGIA	424	90	1,915	28,679	4,981	4,384	2,451
GEORGIA TECH	355	47	801	15,009	1,832	4,285	1,011
GUELPH	88	UA/NA	776	17,019	1,597	1,916	139
HARVARD	413	74	2,433	19,002	5,641	12,048	3,076
HAWAII	149	51	1,226	14,279	2,097	3,215	981
HOUSTON	220	36	1,190	23,897	11,283	4,809	3,059
HOWARD	93	27	1,069	8,075	1,029	1,309	698
ILLINOIS, CHICAGO	210	54	1,456	18,153	4,502	4,219	2,974
ILLINOIS, URBANA	592	88	1,909	37,553	3,134	8,867	2,188
INDIANA	397	88	1,865	33,472	4,486	5,498	2,898

* - Figures were reported on the ARL Questionnaire and have not been verified with the U.S. NCES IPEDS survey
** - Includes both undergraduate and graduate students
UA/NA – Unavailable or Not Applicable

Ph.D., Faculty, and Enrollment Statistics *

	Ph.D.s Awarded	Ph.D. Fields	Faculty	ENROLLMENTS Total FTE Full-time**	Total Part-time**	Graduate FTE Full-time	Graduate Part-time
(Survey Question #) INSTITUTION	(36)	(37)	(38)	(39)	(40)	(41)	(42)
IOWA	289	58	1,137	22,770	5,672	4,891	3,416
IOWA STATE	246	84	1,391	23,094	3,286	2,695	2,046
JOHNS HOPKINS	351	57	1,355	10,115	8,311	4,888	7,577
KANSAS	256	71	1,579	22,849	6,056	4,162	3,400
KENT STATE	125	64	632	18,292	6,055	2,367	2,920
KENTUCKY	233	60	1,198	21,101	4,585	4,506	2,746
LAVAL	236	81	1,432	25,584	18,241	5,592	4,476
LOUISIANA STATE	224	55	1,353	29,754	3,600	4,971	1,547
LOUISVILLE	436	30	1,750	16,053	6,416	6,792	2,538
MCGILL	233	6	1,503	24,703	5,132	5,704	2,108
MCMASTER	145	38	1,152	19,288	3,333	2,231	455
MANITOBA	83	47	1,155	19,968	6,161	2,335	925
MARYLAND	482	70	2,862	35,369	5,464	6,642	3,285
MASSACHUSETTS	224	49	1,147	19,704	4,942	1,981	2,273
MIT	581	36	1,059	9,985	335	5,907	277
MIAMI	108	48	877	13,698	1,552	4,359	787
MICHIGAN	UA/NA	UA/NA	UA/NA	UA/NA	UA/NA	UA/NA	UA/NA
MICHIGAN STATE	425	109	1,939	40,043	6,249	7,837	2,777
MINNESOTA	659	114	1,670	35,736	15,218	9,257	8,981
MISSOURI	274	67	1,114	22,047	3,887	2,514	2,537
MONTREAL	311	79	1,825	37,418	18,490	10,290	3,562
NEBRASKA	234	48	1,159	18,433	3,242	2,665	1,973
NEW MEXICO	214	39	1,683	18,435	8,546	2,766	2,486
NEW YORK	423	71	3,202	29,803	9,605	11,278	7,918
NORTH CAROLINA	459	77	2,844	22,212	4,666	6,514	3,839
NORTH CAROLINA STATE	343	56	1,638	22,711	7,246	3,821	3,382
NORTHWESTERN	366	49	2,491	14,615	3,147	7,675	1,877
NOTRE DAME	133	23	780	11,197	282	2,888	259
OHIO	147	47	1,069	23,723	4,790	2,562	889
OHIO STATE	590	91	2,992	43,796	7,199	10,212	3,274
OKLAHOMA	173	56	1,932	22,826	7,000	3,988	4,046
OKLAHOMA STATE	127	41	1,053	24,245	11,751	2,081	3,902
OREGON	137	42	795	17,103	3,236	2,932	1,049
PENNSYLVANIA	414	69	1,400	19,267	4,038	9,220	2,127
PENNSYLVANIA STATE	606	102	5,067	63,817	11,489	6,925	4,548
PITTSBURGH	372	67	1,846	27,695	6,101	6,515	3,035
PRINCETON	277	46	832	6,708	154	2,030	31
PURDUE	524	57	1,911	34,745	3,908	5,791	2,115
QUEEN'S	137	38	776	16,743	3,698	2,575	378
RICE	137	31	548	4,787	66	1,882	40
ROCHESTER	180	57	1,223	7,136	1,317	2,789	1,103
RUTGERS	342	60	2,152	38,590	11,962	6,079	7,045
SASKATCHEWAN	64	48	998	13,897	3,676	1,810	380

* - Figures were reported on the ARL Questionnaire and have not been verified with the U.S. NCES IPEDS survey
** - Includes both undergraduate and graduate students
UA/NA – Unavailable or Not Applicable

PH.D., FACULTY, AND ENROLLMENT STATISTICS *

	Ph.D.s Awarded	Ph.D. Fields	Faculty	ENROLLMENTS Total FTE Full-time**	Total Part-time**	Graduate FTE Full-time	Graduate Part-time
(Survey Question #) INSTITUTION	(36)	(37)	(38)	(39)	(40)	(41)	(42)
SOUTH CAROLINA	244	59	1,367	21,523	5,300	4,847	3,901
SOUTHERN CALIFORNIA	416	93	3,284	32,581	3,816	16,696	3,118
SOUTHERN ILLINOIS	145	35	869	17,382	4,207	2,350	2,367
SUNY-ALBANY	149	39	637	13,455	3,585	2,244	2,783
SUNY-BUFFALO	380	79	1,201	22,618	4,602	5,707	3,348
SUNY-STONY BROOK	217	38	872	14,035	4,724	2,122	3,610
SYRACUSE	UA/NA	UA/NA	120	UA/NA	UA/NA	1,562	22
TEMPLE	336	50	1,669	28,364	8,422	4,833	5,290
TENNESSEE	454	59	1,844	24,341	3,888	8,516	2,399
TEXAS	701	91	2,467	45,614	4,763	11,726	1,274
TEXAS A&M	528	87	1,902	39,607	5,499	6,559	2,144
TEXAS TECH	166	54	1,683	26,189	4,408	4,797	1,877
TORONTO	631	94	2,657	57,246	8,303	9,813	2,264
TULANE	105	43	1,095	9,989	2,422	4,000	581
UTAH	260	48	1,175	19,977	8,956	4,572	1,586
VANDERBILT	194	44	1,995	10,779	515	4,577	445
VIRGINIA	313	46	1,112	19,258	4,083	5,880	3,332
VIRGINIA TECH	329	47	1,265	24,736	2,883	3,997	2,292
WASHINGTON	452	79	3,952	32,893	6,306	9,660	1,807
WASHINGTON STATE	180	43	1,093	19,298	3,943	2,023	1,205
WASHINGTON U.-ST. LOUIS	168	47	757	10,642	2,568	4,613	1,247
WATERLOO	136	33	816	23,184	2,660	2,285	512
WAYNE STATE	173	56	1,764	18,817	14,320	6,893	5,507
WESTERN ONTARIO	156	48	1,249	23,637	3,589	3,428	435
WISCONSIN	664	109	2,064	33,747	4,259	7,128	1,800
YALE	309	59	1,736	11,144	215	5,902	138
YORK	111	UA/NA	1,357	37,178	7,496	3,121	1,701

SUMMARY DATA: PH.D., FACULTY, AND ENROLLMENT STATISTICS

Median	256	55	1,354	21,664	4,259	4,595	2,121
High	874	137	5,067	63,817	18,490	16,696	8,981
Low	56	6	120	4,787	47	1,023	19
Totals	32,326	6,383	168,129	2,536,614	546,082	568,024	253,605
# of Libraries Responding	111	109	112	111	111	112	112

* - Figures were reported on the ARL Questionnaire and have not been verified with the U.S. NCES IPEDS survey
** - Includes both undergraduate and graduate students
UA/NA - Unavailable or Not Applicable

Rank Order Tables of University Libraries, 2004-05

Summary of Rank Order Tables for University Libraries, 2004-05

The table below presents the rank for each university library in each of the eighteen categories for which rank order tables are prepared. The table numbers in the chart below refer to the data categories listed below. The number of libraries indicates the number of ARL university libraries supplying data in each category.

TABLE	DATA CATEGORY	NO. OF LIBS.	TABLE	DATA CATEGORY	NO. OF LIBS.
1	VOLUMES IN LIBRARY	113	11	EXPENDITURES FOR MONOGRAPHS	112
2	VOLUMES ADDED (GROSS)	113	12	SERIALS PURCHASED (SUBS.)	89
3	CURRENT SERIALS (TOTALS)	113	13	EXPENDITURES FOR SERIALS	112
4	MICROFORM HOLDINGS	112	14	TOTAL ITEMS LOANED (ILL)	111
5	GOVERNMENT DOCUMENTS	99	15	TOTAL ITEMS BORROWED (ILL/DD)	111
6	MATERIALS EXPENDITURES	113	16	PROFESSIONAL STAFF (FTE)	113
7	SALARIES & WAGES EXPENDITURES	112	17	SUPPORT STAFF (FTE)	113
8	OTHER OPERATING EXPENDITURES	112	18	TOTAL STAFF (FTE)	113
9	TOTAL LIBRARY EXPENDITURES	113	19	EXPENDITURES FOR E-MATERIALS	113
10	MONOGRAPHS PURCHASED (VOLS.)	101	20	E-MATERIALS AS % OF TOTAL MATERIALS	113

INSTITUTION	TABLES																			
	1	2	3	4	5	6	7	8	9	10	11	12	13	14	15	16	17	18	19	20
ALABAMA	83	66	71	63	11	92	97	77	98	22	60	49	89	107	103	82	94	81	68	40
ALBERTA	17	8	47	74	73	19	28	93	34	UA	11	36	22	4	34	56	13	38	4	16
ARIZONA	23	28	67	24	UA	31	32	23	30	24	32	UA	23	30	46	39	37	33	11	33
ARIZONA STATE	34	49	69	7	16	44	33	40	41	13	44	55	39	15	54	51	18	33	19	28
AUBURN	71	99	78	94	17	109	110	51	107	75	103	85	109	106	105	97	108	108	88	42
BOSTON	94	84	64	48	UA	65	63	66	66	91	105	UA	49	95	82	45	74	41	106	103
BOSTON COLLEGE	106	70	26	61	49	67	71	103	73	UA	66	21	59	91	98	60	96	81	92	84
BRIGHAM YOUNG	40	32	83	72	33	37	40	34	38	29	47	68	16	59	23	30	108	24	93	101
BRITISH COLUMBIA	22	33	25	41	UA	28	27	50	31	UA	52	UA	12	77	65	41	21	32	9	27
BROWN	45	79	72	104	73	64	62	76	67	50	37	52	73	55	33	56	76	76	48	36
CALIFORNIA, BERKELEY	5	5	3	12	55	7	4	3	4	UA	3	UA	17	35	66	5	24	5	UA	UA
CALIFORNIA, DAVIS	46	64	50	58	57	88	52	81	69	UA	50	UA	108	62	28	77	41	55	55	19
CALIFORNIA, IRVINE	91	87	90	91	52	80	60	68	71	58	57	57	81	67	50	79	55	65	49	13
CALIFORNIA, LOS ANGELES	9	22	7	27	58	25	5	4	8	10	9	UA	42	13	19	8	8	6	59	87
CALIFORNIA, RIVERSIDE	98	58	82	106	44	103	96	88	103	42	69	64	103	82	85	108	87	102	94	58

UA – Unavailable

INSTITUTION	TABLES																			
	1	2	3	4	5	6	7	8	9	10	11	12	13	14	15	16	17	18	19	20
CALIFORNIA, SAN DIEGO	61	24	73	83	73	62	25	19	33	30	30	58	69	21	49	47	25	30	70	76
CALIFORNIA, SANTA BARBARA	73	112	52	73	34	102	70	33	70	62	48	51	105	79	35	102	47	70	99	75
CASE WESTERN RESERVE	88	110	85	99	63	82	94	63	87	97	108	80	65	47	21	75	110	107	74	52
CHICAGO	12	9	42	89	UA	12	37	41	29	4	10	UA	10	33	59	73	28	39	39	83
CINCINNATI	62	30	57	81	61	40	50	60	49	32	26	22	50	19	15	31	97	66	17	23
COLORADO	43	41	81	13	10	56	72	78	65	40	68	86	46	27	47	97	60	74	23	14
COLORADO STATE	108	53	80	100	45	93	102	73	97	57	67	62	94	31	6	106	107	109	58	20
COLUMBIA	6	4	4	26	50	5	6	7	5	1	5	7	1	23	12	3	10	7	2	63
CONNECTICUT	56	96	30	53	66	63	26	79	43	71	93	31	43	24	31	58	93	72	37	25
CORNELL	11	14	6	4	UA	11	9	6	9	UA	7	3	26	14	38	24	5	11	12	61
DARTMOUTH	86	90	32	98	73	76	75	70	74	69	61	13	64	60	44	86	61	85	36	6
DELAWARE	77	86	111	79	31	74	78	90	82	60	45	88	77	94	74	88	71	84	21	5
DUKE	21	47	70	56	5	24	21	35	25	14	23	26	13	45	73	11	40	31	40	78
EMORY	63	31	58	38	39	13	31	21	24	35	12	UA	67	40	83	46	35	42	47	85
FLORIDA	35	55	10	6	6	35	29	45	36	25	41	1	20	89	58	28	22	16	24	45
FLORIDA STATE	72	103	46	5	15	81	100	39	83	88	82	18	92	37	75	68	52	64	89	73
GEORGE WASHINGTON	104	71	105	92	73	51	57	69	57	77	24	72	72	53	10	62	57	63	87	91
GEORGETOWN	87	43	79	67	65	49	45	48	47	44	65	45	40	52	67	55	51	57	84	92
GEORGIA	32	45	33	16	9	33	54	58	46	26	43	33	27	26	104	58	25	39	13	35
GEORGIA TECH	95	97	74	50	73	107	111	100	111	100	111	79	106	110	110	103	101	112	UA	UA
GUELPH	113	113	112	102	73	111	109	82	110	95	95	82	111	34	5	109	100	111	60	4
HARVARD	1	1	2	2	UA	2	1	1	1	UA	2	UA	2	38	93	1	1	1	25	105
HAWAII	49	73	89	29	73	98	77	94	89	83	85	63	86	111	107	66	90	85	80	51
HOUSTON	99	88	100	49	20	69	91	42	79	54	81	67	61	66	81	54	103	90	105	102
HOWARD	97	107	113	75	72	112	103	107	112	101	112	87	107	98	109	93	97	100	102	47
ILLINOIS, CHICAGO	100	75	88	65	18	68	68	96	72	51	92	40	52	10	22	65	53	51	61	53
ILLINOIS, URBANA	3	7	5	3	UA	17	11	31	16	6	29	UA	5	5	2	6	20	12	52	89
INDIANA	13	15	11	44	36	20	23	29	22	12	13	4	51	7	18	9	27	15	30	70
IOWA	31	21	29	11	23	26	41	80	40	11	33	28	33	18	53	53	44	43	33	69

UA – Unavailable

INSTITUTION	TABLES																			
	1	2	3	4	5	6	7	8	9	10	11	12	13	14	15	16	17	18	19	20
IOWA STATE	90	83	77	78	73	59	80	99	75	59	56	70	47	58	87	97	82	94	43	38
JOHNS HOPKINS	39	72	9	54	70	23	34	18	27	55	71	12	3	48	26	18	33	36	15	48
KANSAS	36	35	27	77	19	61	59	57	60	39	55	37	62	28	25	35	67	45	98	97
KENT STATE	76	78	109	111	73	113	112	84	113	99	110	83	112	103	106	113	112	113	103	31
KENTUCKY	53	42	76	20	7	48	66	65	59	28	73	60	37	56	45	43	58	50	66	82
LAVAL	75	27	91	112	73	66	69	112	78	52	70	53	54	90	89	82	39	71	UA	UA
LOUISIANA STATE	50	67	17	15	25	79	98	95	96	8	39	UA	91	109	51	97	77	91	108	107
LOUISVILLE	109	36	95	103	73	58	90	67	76	41	83	48	32	78	100	97	85	93	91	88
MCGILL	41	50	31	105	22	30	49	32	37	34	28	14	44	93	57	70	32	52	6	10
MCMASTER	111	104	86	108	73	96	105	108	109	27	88	74	93	97	90	112	72	102	57	8
MANITOBA	105	100	93	110	48	91	65	92	81	33	76	61	96	72	24	79	45	68	63	32
MARYLAND	54	34	68	32	51	60	44	44	51	16	63	77	101	96	20	21	87	56	28	18
MASSACHUSETTS	58	101	51	97	41	108	81	89	99	UA	100	24	99	43	32	92	97	97	96	59
MIT	74	92	96	101	UA	77	39	53	55	87	84	78	60	105	101	44	81	81	76	65
MIAMI	82	39	54	64	73	36	73	55	56	43	40	UA	34	61	84	47	62	53	65	86
MICHIGAN	8	6	1	UA	UA	3	UA	UA	6	UA	UA	UA	UA	UA	UA	10	5	8	UA	UA
MICHIGAN STATE	27	62	55	17	73	47	46	46	45	47	49	54	31	32	39	62	55	46	44	49
MINNESOTA	15	11	43	14	2	21	17	10	17	49	34	30	7	1	29	37	13	20	20	62
MISSOURI	55	77	59	9	3	99	88	56	91	68	98	25	84	16	17	90	62	87	82	50
MONTREAL	65	37	40	109	73	46	24	97	39	18	35	20	36	70	79	32	8	23	34	39
NEBRASKA	70	61	38	46	27	85	87	59	86	93	101	23	70	69	91	96	64	79	78	64
NEW MEXICO	78	69	110	45	26	100	42	52	62	74	107	81	90	88	69	66	34	49	95	66
NEW YORK	26	25	21	21	30	9	14	14	12	19	19	17	4	100	60	24	12	17	10	56
NORTH CAROLINA	20	20	23	43	4	18	19	27	20	17	27	34	24	22	92	15	19	18	100	108
NORTH CAROLINA STATE	44	13	18	35	73	54	47	20	42	72	59	47	57	102	55	23	68	47	86	90
NORTHWESTERN	30	29	49	52	67	29	30	47	35	36	21	UA	25	63	27	28	45	33	8	22
NOTRE DAME	59	51	97	71	UA	57	58	74	61	56	25	UA	75	83	88	86	36	61	35	26
OHIO	81	68	104	82	62	104	92	43	94	31	91	71	104	6	4	73	103	95	81	41
OHIO STATE	18	18	62	31	68	52	22	30	32	9	38	59	82	2	1	19	37	19	18	9

UA – Unavailable

INSTITUTION	TABLES																			
	1	2	3	4	5	6	7	8	9	10	11	12	13	14	15	16	17	18	19	20
OKLAHOMA	28	59	36	55	64	32	95	85	64	65	17	27	68	65	30	79	84	72	77	98
OKLAHOMA STATE	80	40	15	51	53	84	107	64	95	73	75	UA	100	51	77	75	95	76	46	7
OREGON	79	105	106	90	29	106	79	104	100	81	80	76	110	9	8	95	77	75	101	71
PENNSYLVANIA	19	26	39	62	UA	15	20	26	21	UA	15	16	8	44	3	24	29	21	31	77
PENNSYLVANIA STATE	25	38	12	39	73	6	7	13	7	15	18	8	56	20	11	12	2	4	1	30
PITTSBURGH	29	10	41	28	73	16	36	36	28	UA	22	UA	14	11	68	27	31	28	14	55
PRINCETON	16	16	14	19	54	10	16	25	15	5	6	5	38	57	37	14	17	22	51	93
PURDUE	85	80	101	86	28	45	53	49	52	78	74	75	41	68	52	70	49	60	83	94
QUEEN'S	93	108	99	68	8	83	93	75	90	79	90	56	71	74	95	107	66	98	50	11
RICE	92	91	107	84	60	53	108	86	85	46	16	84	79	104	96	88	106	110	71	81
ROCHESTER	42	74	92	42	32	87	67	71	77	82	96	UA	74	73	70	47	105	87	107	100
RUTGERS	33	46	28	30	1	41	12	72	23	76	78	9	28	76	56	34	11	14	45	60
SASKATCHEWAN	112	63	94	87	35	89	104	106	104	80	62	65	88	75	63	104	77	106	72	46
SOUTH CAROLINA	47	95	22	40	12	90	74	22	68	66	87	UA	76	87	72	68	68	54	62	29
SOUTHERN CALIFORNIA	37	60	56	25	46	27	18	12	18	48	51	32	35	81	102	22	54	25	38	72
SOUTHERN ILLINOIS	68	94	44	47	42	78	84	102	88	92	99	43	63	25	36	90	82	79	69	57
SUNY-ALBANY	107	102	48	93	73	110	85	105	108	96	104	19	102	108	78	85	101	99	104	79
SUNY-BUFFALO	51	82	61	36	UA	73	48	87	63	94	94	50	58	50	76	17	111	57	42	15
SUNY-STONY BROOK	103	109	63	70	40	94	86	111	102	98	109	38	80	84	94	51	113	104	79	54
SYRACUSE	60	89	102	10	43	105	76	98	92	85	97	69	97	99	64	72	48	68	73	21
TEMPLE	67	93	84	85	56	75	82	91	84	64	89	66	87	85	86	77	85	78	41	12
TENNESSEE	69	57	60	59	71	38	51	62	50	61	53	41	18	49	42	47	42	48	85	99
TEXAS	7	12	34	18	73	14	10	15	11	23	14	39	21	29	41	12	7	10	22	68
TEXAS A&M	48	23	19	34	59	8	43	37	26	7	8	10	11	36	9	20	42	27	3	24
TEXAS TECH	89	56	45	96	69	50	64	38	54	38	42	44	48	64	40	39	58	37	90	95
TORONTO	4	3	13	37	UA	4	3	9	3	3	4	6	6	54	99	7	4	3	54	104
TULANE	96	76	108	95	24	86	99	101	101	53	72	73	83	80	97	93	72	92	97	80
UTAH	57	52	53	76	14	72	38	24	44	UA	54	UA	66	42	61	60	23	29	109	109
VANDERBILT	66	48	66	88	47	43	55	54	53	70	64	42	15	71	80	35	68	61	29	44

UA – Unavailable

INSTITUTION	TABLES																			
	1	2	3	4	5	6	7	8	9	10	11	12	13	14	15	16	17	18	19	20
VIRGINIA	24	44	24	33	21	39	15	11	19	37	46	35	45	41	62	37	13	26	53	74
VIRGINIA TECH	102	98	65	22	37	97	106	83	105	84	86	29	95	86	71	110	87	104	67	34
WASHINGTON	14	19	20	8	73	22	13	17	14	20	20	UA	9	8	14	16	13	13	27	67
WASHINGTON STATE	101	54	75	57	UA	101	89	61	93	89	102	UA	85	UA	UA	104	92	100	64	17
WASHINGTON U.-ST. LOUIS	38	65	37	80	73	42	35	2	13	67	77	UA	29	46	13	33	49	43	26	37
WATERLOO	110	111	103	107	73	95	101	109	106	90	106	UA	98	12	16	111	77	96	32	2
WAYNE STATE	52	85	98	69	38	70	56	16	48	86	79	UA	55	17	43	41	90	57	5	1
WESTERN ONTARIO	64	106	35	66	73	55	83	110	80	63	58	15	30	92	108	64	74	89	7	3
WISCONSIN	10	17	16	23	13	34	8	8	10	21	31	11	53	3	7	4	29	9	75	96
YALE	2	2	8	1	73	1	2	5	2	2	1	2	19	39	48	2	3	2	16	106
YORK	84	81	87	60	73	71	61	28	58	45	36	46	78	101	111	82	65	67	56	43

UA – Unavailable

Rank Order Table 1
Volumes in Library

Rank	Institution	Value	Rank	Institution	Value
1	HARVARD	15,555,533	58	MASSACHUSETTS	3,204,025
2	YALE	12,025,695	59	NOTRE DAME	3,185,926
3	ILLINOIS, URBANA	10,370,777	60	SYRACUSE	3,161,529
4	TORONTO	10,342,574	61	CALIF., SAN DIEGO	3,149,836
5	CALIF., BERKELEY	9,985,905	62	CINCINNATI	3,123,318
6	COLUMBIA	9,277,042	63	EMORY	3,107,525
7	TEXAS	8,937,002	64	WESTERN ONTARIO	3,085,319
8	MICHIGAN	8,133,917	65	MONTREAL	3,057,166
9	CALIF., LOS ANGELES	8,064,896	66	VANDERBILT	3,056,707
10	WISCONSIN	7,911,834	67	TEMPLE	3,016,007
11	CORNELL	7,644,371	68	SOUTHERN ILLINOIS	2,999,736
12	CHICAGO	7,363,549	69	TENNESSEE	2,971,837
13	INDIANA	7,241,929	70	NEBRASKA	2,942,946
14	WASHINGTON	6,639,850	71	AUBURN	2,918,859
15	MINNESOTA	6,587,430	72	FLORIDA STATE	2,889,810
16	PRINCETON	6,495,597	73	CALIF., SANTA BARBARA	2,837,752
17	ALBERTA	6,190,396	74	MIT	2,782,406
18	OHIO STATE	5,936,434	75	LAVAL	2,731,341
19	PENNSYLVANIA	5,760,065	76	KENT STATE	2,715,986
20	NORTH CAROLINA	5,710,686	77	DELAWARE	2,667,242
21	DUKE	5,560,966	78	NEW MEXICO	2,666,970
22	BRITISH COLUMBIA	5,523,194	79	OREGON	2,643,035
23	ARIZONA	5,340,726	80	OKLAHOMA STATE	2,624,680
24	VIRGINIA	5,053,162	81	OHIO	2,599,791
25	PENNSYLVANIA STATE	5,031,196	82	MIAMI	2,571,045
26	NEW YORK	4,939,724	83	ALABAMA	2,518,290
27	MICHIGAN STATE	4,830,861	84	YORK	2,489,193
28	OKLAHOMA	4,796,089	85	PURDUE	2,475,242
29	PITTSBURGH	4,786,175	86	DARTMOUTH	2,474,288
30	NORTHWESTERN	4,603,824	87	GEORGETOWN	2,473,208
31	IOWA	4,551,217	88	CASE WESTERN RESERVE	2,471,504
32	GEORGIA	4,179,316	89	TEXAS TECH	2,465,796
33	RUTGERS	4,169,347	90	IOWA STATE	2,444,263
34	ARIZONA STATE	4,156,732	91	CALIF., IRVINE	2,439,525
35	FLORIDA	4,137,528	92	RICE	2,437,177
36	KANSAS	4,121,573	93	QUEEN'S	2,437,168
37	SOUTHERN CALIFORNIA	3,921,704	94	BOSTON	2,427,253
38	WASHINGTON U.-ST. LOUIS	3,694,504	95	GEORGIA TECH	2,410,964
39	JOHNS HOPKINS	3,648,821	96	TULANE	2,403,728
40	BRIGHAM YOUNG	3,621,577	97	HOWARD	2,388,073
41	MCGILL	3,568,651	98	CALIF., RIVERSIDE	2,368,843
42	ROCHESTER	3,564,683	99	HOUSTON	2,357,009
43	COLORADO	3,554,826	100	ILLINOIS, CHICAGO	2,284,649
44	NORTH CAROLINA STATE	3,530,949	101	WASHINGTON STATE	2,256,229
45	BROWN	3,509,710	102	VIRGINIA TECH	2,237,106
46	CALIF., DAVIS	3,490,356	103	SUNY-STONY BROOK	2,216,589
47	SOUTH CAROLINA	3,478,256	104	GEORGE WASHINGTON	2,167,871
48	TEXAS A&M	3,437,376	105	MANITOBA	2,158,148
49	HAWAII	3,410,468	106	BOSTON COLLEGE	2,124,242
50	LOUISIANA STATE	3,406,434	107	SUNY-ALBANY	2,094,048
51	SUNY-BUFFALO	3,390,583	108	COLORADO STATE	2,026,508
52	WAYNE STATE	3,383,826	109	LOUISVILLE	2,015,752
53	KENTUCKY	3,286,731	110	WATERLOO	2,006,887
54	MARYLAND	3,259,600	111	MCMASTER	1,990,867
55	MISSOURI	3,249,783	112	SASKATCHEWAN	1,976,606
56	CONNECTICUT	3,247,976	113	GUELPH	1,559,078
57	UTAH	3,230,854			

Rank Order Table 2
Volumes Added (Gross)

Rank	Institution	Value	Rank	Institution	Value
1	HARVARD	429,344	58	CALIFORNIA, RIVERSIDE	63,958
2	YALE	302,604	59	OKLAHOMA	63,849
3	TORONTO	269,669	60	SOUTHERN CALIFORNIA	63,817
4	COLUMBIA	223,923	61	NEBRASKA	61,649
5	CALIFORNIA, BERKELEY	192,944	62	MICHIGAN STATE	60,188
6	MICHIGAN	189,373	63	SASKATCHEWAN	57,844
7	ILLINOIS, URBANA	179,898	64	CALIFORNIA, DAVIS	57,786
8	ALBERTA	166,508	65	WASHINGTON U.-ST. LOUIS	57,710
9	CHICAGO	165,351	66	ALABAMA	57,476
10	PITTSBURGH	153,774	67	LOUISIANA STATE	57,259
11	MINNESOTA	148,474	68	OHIO	57,126
12	TEXAS	147,811	69	NEW MEXICO	56,162
13	NORTH CAROLINA STATE	147,268	70	BOSTON COLLEGE	55,116
14	CORNELL	136,758	71	GEORGE WASHINGTON	54,646
15	INDIANA	133,295	72	JOHNS HOPKINS	54,597
16	PRINCETON	127,750	73	HAWAII	54,437
17	WISCONSIN	124,409	74	ROCHESTER	53,544
18	OHIO STATE	124,288	75	ILLINOIS, CHICAGO	53,000
19	WASHINGTON	120,914	76	TULANE	51,814
20	NORTH CAROLINA	119,474	77	MISSOURI	51,179
21	IOWA	118,346	78	KENT STATE	50,327
22	CALIFORNIA, LOS ANGELES	110,858	79	BROWN	49,800
23	TEXAS A&M	110,204	80	PURDUE	49,636
24	CALIFORNIA, SAN DIEGO	109,064	81	YORK	48,948
25	NEW YORK	108,920	82	SUNY-BUFFALO	48,556
26	PENNSYLVANIA	105,875	83	IOWA STATE	48,171
27	LAVAL	103,857	84	BOSTON	47,070
28	ARIZONA	103,576	85	WAYNE STATE	46,707
29	NORTHWESTERN	89,205	86	DELAWARE	46,237
30	CINCINNATI	87,762	87	CALIFORNIA, IRVINE	46,193
31	EMORY	87,371	88	HOUSTON	44,841
32	BRIGHAM YOUNG	86,125	89	SYRACUSE	44,840
33	BRITISH COLUMBIA	85,516	90	DARTMOUTH	44,517
34	MARYLAND	84,206	91	RICE	44,338
35	KANSAS	83,469	92	MIT	44,256
36	LOUISVILLE	83,157	93	TEMPLE	44,143
37	MONTREAL	81,449	94	SOUTHERN ILLINOIS	43,765
38	PENNSYLVANIA STATE	81,008	95	SOUTH CAROLINA	42,928
39	MIAMI	80,204	96	CONNECTICUT	42,220
40	OKLAHOMA STATE	78,575	97	GEORGIA TECH	40,950
41	COLORADO	78,213	98	VIRGINIA TECH	38,048
42	KENTUCKY	77,597	99	AUBURN	37,370
43	GEORGETOWN	77,272	100	MANITOBA	37,089
44	VIRGINIA	77,249	101	MASSACHUSETTS	35,823
45	GEORGIA	77,062	102	SUNY-ALBANY	35,707
46	RUTGERS	76,198	103	FLORIDA STATE	33,874
47	DUKE	73,037	104	MCMASTER	33,375
48	VANDERBILT	72,520	105	OREGON	33,110
49	ARIZONA STATE	72,240	106	WESTERN ONTARIO	32,907
50	MCGILL	68,311	107	HOWARD	32,345
51	NOTRE DAME	67,696	108	QUEEN'S	31,930
52	UTAH	67,556	109	SUNY-STONY BROOK	26,304
53	COLORADO STATE	67,518	110	CASE WESTERN RESERVE	26,268
54	WASHINGTON STATE	64,702	111	WATERLOO	25,501
55	FLORIDA	64,670	112	CALIFORNIA, SANTA BARBARA	21,565
56	TEXAS TECH	64,585	113	GUELPH	19,453
57	TENNESSEE	64,013			

RANK ORDER TABLE 3
CURRENT SERIALS (TOTALS)

RANK	INSTITUTION	VALUE	RANK	INSTITUTION	VALUE
1	MICHIGAN	124,809	58	EMORY	36,898
2	HARVARD	98,773	59	MISSOURI	36,502
3	CALIFORNIA, BERKELEY	93,256	60	TENNESSEE	36,231
4	COLUMBIA	89,452	61	SUNY-BUFFALO	34,965
5	ILLINOIS, URBANA	88,650	62	OHIO STATE	34,745
6	CORNELL	80,633	63	SUNY-STONY BROOK	34,428
7	CALIFORNIA, LOS ANGELES	78,236	64	BOSTON	33,983
8	YALE	76,022	65	VIRGINIA TECH	33,925
9	JOHNS HOPKINS	73,872	66	VANDERBILT	33,883
10	FLORIDA	70,882	67	ARIZONA	33,575
11	INDIANA	70,672	68	MARYLAND	33,477
12	PENNSYLVANIA STATE	68,445	69	ARIZONA STATE	33,122
13	TORONTO	67,675	70	DUKE	32,963
14	PRINCETON	63,987	71	ALABAMA	32,796
15	OKLAHOMA STATE	61,507	72	BROWN	32,632
16	WISCONSIN	60,457	73	CALIFORNIA, SAN DIEGO	31,478
17	LOUISIANA STATE	58,461	74	GEORGIA TECH	30,799
18	NORTH CAROLINA STATE	57,486	75	WASHINGTON STATE	30,648
19	TEXAS A&M	56,903	76	KENTUCKY	30,250
20	WASHINGTON	55,932	77	IOWA STATE	29,850
21	NEW YORK	54,024	78	AUBURN	29,355
22	SOUTH CAROLINA	53,610	79	GEORGETOWN	29,254
23	NORTH CAROLINA	53,444	80	COLORADO STATE	28,837
24	VIRGINIA	53,370	81	COLORADO	28,310
25	BRITISH COLUMBIA	53,114	82	CALIFORNIA, RIVERSIDE	28,151
26	BOSTON COLLEGE	52,338	83	BRIGHAM YOUNG	27,805
27	KANSAS	50,992	84	TEMPLE	26,979
28	RUTGERS	50,229	85	CASE WESTERN RESERVE	26,965
29	IOWA	50,214	86	MCMASTER	26,908
30	CONNECTICUT	50,066	87	YORK	26,838
31	MCGILL	49,433	88	ILLINOIS, CHICAGO	26,665
32	DARTMOUTH	48,366	89	HAWAII	26,605
33	GEORGIA	48,227	90	CALIFORNIA, IRVINE	26,493
34	TEXAS	46,857	91	LAVAL	26,389
35	WESTERN ONTARIO	45,424	92	ROCHESTER	26,218
36	OKLAHOMA	44,975	93	MANITOBA	26,039
37	WASHINGTON U.-ST. LOUIS	44,806	94	SASKATCHEWAN	25,881
38	NEBRASKA	44,105	95	LOUISVILLE	23,575
39	PENNSYLVANIA	43,931	96	MIT	23,038
40	MONTREAL	43,707	97	NOTRE DAME	22,866
41	PITTSBURGH	43,441	98	WAYNE STATE	22,606
42	CHICAGO	43,396	99	QUEEN'S	21,921
43	MINNESOTA	43,303	100	HOUSTON	21,845
44	SOUTHERN ILLINOIS	43,083	101	PURDUE	21,598
45	TEXAS TECH	42,860	102	SYRACUSE	20,637
46	FLORIDA STATE	42,076	103	WATERLOO	20,028
47	ALBERTA	41,947	104	OHIO	19,928
48	SUNY-ALBANY	41,707	105	GEORGE WASHINGTON	18,909
49	NORTHWESTERN	41,693	106	OREGON	18,826
50	CALIFORNIA, DAVIS	41,273	107	RICE	16,954
51	MASSACHUSETTS	41,207	108	TULANE	16,588
52	CALIFORNIA, SANTA BARBARA	40,459	109	KENT STATE	13,357
53	UTAH	40,446	110	NEW MEXICO	13,337
54	MIAMI	38,432	111	DELAWARE	12,530
55	MICHIGAN STATE	37,832	112	GUELPH	11,817
56	SOUTHERN CALIFORNIA	37,809	113	HOWARD	11,317
57	CINCINNATI	37,668			

Rank Order Table 4
Microform Units

Rank	Institution	Value	Rank	Institution	Value
1	YALE	10,037,556	58	CALIFORNIA, DAVIS	4,203,651
2	HARVARD	9,903,096	59	TENNESSEE	4,168,542
3	ILLINOIS, URBANA	9,302,770	60	YORK	4,101,547
4	CORNELL	8,204,567	61	BOSTON COLLEGE	4,050,845
5	FLORIDA STATE	8,057,015	62	PENNSYLVANIA	4,048,727
6	FLORIDA	7,680,340	63	ALABAMA	4,014,614
7	ARIZONA STATE	7,602,664	64	MIAMI	3,924,529
8	WASHINGTON	7,588,429	65	ILLINOIS, CHICAGO	3,903,703
9	MISSOURI	7,437,324	66	WESTERN ONTARIO	3,883,392
10	SYRACUSE	7,334,236	67	GEORGETOWN	3,877,478
11	IOWA	7,131,140	68	QUEEN'S	3,861,728
12	CALIFORNIA, BERKELEY	6,825,994	69	WAYNE STATE	3,847,036
13	COLORADO	6,802,839	70	SUNY-STONY BROOK	3,832,124
14	MINNESOTA	6,715,389	71	NOTRE DAME	3,831,274
15	LOUISIANA STATE	6,697,824	72	BRIGHAM YOUNG	3,825,198
16	GEORGIA	6,537,902	73	CALIFORNIA, SANTA BARBARA	3,759,140
17	MICHIGAN STATE	6,518,542	74	ALBERTA	3,716,844
18	TEXAS	6,510,523	75	HOWARD	3,671,886
19	PRINCETON	6,424,007	76	UTAH	3,608,147
20	KENTUCKY	6,347,361	77	KANSAS	3,606,499
21	NEW YORK	6,332,917	78	IOWA STATE	3,473,037
22	VIRGINIA TECH	6,292,655	79	DELAWARE	3,419,588
23	WISCONSIN	6,248,997	80	WASHINGTON U.-ST. LOUIS	3,355,312
24	ARIZONA	6,210,417	81	CINCINNATI	3,340,167
25	SOUTHERN CALIFORNIA	6,187,918	82	OHIO	3,206,766
26	COLUMBIA	6,176,325	83	CALIFORNIA, SAN DIEGO	3,201,291
27	CALIFORNIA, LOS ANGELES	6,166,293	84	RICE	3,168,578
28	PITTSBURGH	6,145,754	85	TEMPLE	3,147,750
29	HAWAII	6,001,718	86	PURDUE	3,145,402
30	RUTGERS	5,906,846	87	SASKATCHEWAN	3,069,886
31	OHIO STATE	5,834,511	88	VANDERBILT	3,022,373
32	MARYLAND	5,736,084	89	CHICAGO	3,011,073
33	VIRGINIA	5,541,742	90	OREGON	2,909,868
34	TEXAS A&M	5,480,289	91	CALIFORNIA, IRVINE	2,884,821
35	NORTH CAROLINA STATE	5,413,654	92	GEORGE WASHINGTON	2,864,551
36	SUNY-BUFFALO	5,406,490	93	SUNY-ALBANY	2,860,553
37	TORONTO	5,371,934	94	AUBURN	2,670,252
38	EMORY	5,368,365	95	TULANE	2,652,736
39	PENNSYLVANIA STATE	5,271,498	96	TEXAS TECH	2,591,561
40	SOUTH CAROLINA	5,210,142	97	MASSACHUSETTS	2,575,821
41	BRITISH COLUMBIA	5,117,219	98	DARTMOUTH	2,567,627
42	ROCHESTER	5,081,610	99	CASE WESTERN RESERVE	2,548,156
43	NORTH CAROLINA	4,987,306	100	COLORADO STATE	2,520,094
44	INDIANA	4,978,205	101	MIT	2,391,503
45	NEW MEXICO	4,690,017	102	GUELPH	2,305,363
46	NEBRASKA	4,618,411	103	LOUISVILLE	2,190,246
47	SOUTHERN ILLINOIS	4,566,453	104	BROWN	1,928,655
48	BOSTON	4,557,897	105	MCGILL	1,793,955
49	HOUSTON	4,531,918	106	CALIFORNIA, RIVERSIDE	1,772,053
50	GEORGIA TECH	4,495,012	107	WATERLOO	1,707,697
51	OKLAHOMA STATE	4,488,446	108	MCMASTER	1,597,371
52	NORTHWESTERN	4,405,025	109	MONTREAL	1,516,331
53	CONNECTICUT	4,397,856	110	MANITOBA	1,468,054
54	JOHNS HOPKINS	4,380,319	111	KENT STATE	1,369,182
55	OKLAHOMA	4,316,009	112	LAVAL	1,312,518
56	DUKE	4,230,496	113	MICHIGAN	UA
57	WASHINGTON STATE	4,224,734			

UA – Unavailable

RANK ORDER TABLE 5
GOVERNMENT DOCUMENTS

RANK	INSTITUTION	VALUE	RANK	INSTITUTION	VALUE
1	RUTGERS	2,661,336	58	CALIFORNIA, LOS ANGELES	69,577
2	MINNESOTA	2,652,842	59	TEXAS A&M	65,519
3	MISSOURI	1,663,610	60	RICE	55,896
4	NORTH CAROLINA	1,625,932	61	CINCINNATI	38,528
5	DUKE	1,246,974	62	OHIO	31,366
6	FLORIDA	1,223,826	63	CASE WESTERN RESERVE	24,675
7	KENTUCKY	1,125,174	64	OKLAHOMA	18,480
8	QUEEN'S	1,099,003	65	GEORGETOWN	9,385
9	GEORGIA	1,055,763	66	CONNECTICUT	7,922
10	COLORADO	973,069	67	NORTHWESTERN	7,808
11	ALABAMA	930,419	68	OHIO STATE	6,287
12	SOUTH CAROLINA	920,317	69	TEXAS TECH	4,222
13	WISCONSIN	909,069	70	JOHNS HOPKINS	3,897
14	UTAH	885,549	71	TENNESSEE	1,776
15	FLORIDA STATE	852,152	72	HOWARD	540
16	ARIZONA STATE	797,044	73	ALBERTA	0
17	AUBURN	778,424	74	BROWN	0
18	ILLINOIS, CHICAGO	762,943	75	CALIFORNIA, SAN DIEGO	0
19	KANSAS	732,058	76	DARTMOUTH	0
20	HOUSTON	680,971	77	GEORGE WASHINGTON	0
21	VIRGINIA	676,334	78	GEORGIA TECH	0
22	MCGILL	668,033	79	GUELPH	0
23	IOWA	665,618	80	HAWAII	0
24	TULANE	616,614	81	IOWA STATE	0
25	LOUISIANA STATE	613,533	82	KENT STATE	0
26	NEW MEXICO	609,773	83	LAVAL	0
27	NEBRASKA	552,672	84	LOUISVILLE	0
28	PURDUE	488,674	85	MCMASTER	0
29	OREGON	467,921	86	MIAMI	0
30	NEW YORK	450,565	87	MICHIGAN STATE	0
31	DELAWARE	445,614	88	MONTREAL	0
32	ROCHESTER	443,509	89	NORTH CAROLINA STATE	0
33	BRIGHAM YOUNG	414,836	90	PENNSYLVANIA STATE	0
34	CALIFORNIA, SANTA BARBARA	413,314	91	PITTSBURGH	0
35	SASKATCHEWAN	412,224	92	SUNY-ALBANY	0
36	INDIANA	405,323	93	TEXAS	0
37	VIRGINIA TECH	399,525	94	WASHINGTON	0
38	WAYNE STATE	388,182	95	WASHINGTON U.-ST. LOUIS	0
39	EMORY	352,310	96	WATERLOO	0
40	SUNY-STONY BROOK	335,159	97	WESTERN ONTARIO	0
41	MASSACHUSETTS	315,238	98	YALE	0
42	SOUTHERN ILLINOIS	313,173	99	YORK	0
43	SYRACUSE	306,460	100	ARIZONA	UA
44	CALIFORNIA, RIVERSIDE	297,740	101	BOSTON	UA
45	COLORADO STATE	286,420	102	BRITISH COLUMBIA	UA
46	SOUTHERN CALIFORNIA	266,024	103	CHICAGO	UA
47	VANDERBILT	255,423	104	CORNELL	UA
48	MANITOBA	225,236	105	HARVARD	UA
49	BOSTON COLLEGE	214,674	106	ILLINOIS, URBANA	UA
50	COLUMBIA	193,534	107	MIT	UA
51	MARYLAND	185,837	108	MICHIGAN	UA
52	CALIFORNIA, IRVINE	170,730	109	NOTRE DAME	UA
53	OKLAHOMA STATE	145,996	110	PENNSYLVANIA	UA
54	PRINCETON	123,609	111	SUNY-BUFFALO	UA
55	CALIFORNIA, BERKELEY	98,774	112	TORONTO	UA
56	TEMPLE	87,621	113	WASHINGTON STATE	UA
57	CALIFORNIA, DAVIS	74,595			

UA – Unavailable

Rank Order Table 6
Total Library Materials Expenditures

Rank	Institution	Value	Rank	Institution	Value
1	YALE	31,194,924	58	LOUISVILLE	8,641,096
2	HARVARD	27,569,823	59	IOWA STATE	8,599,786
3	MICHIGAN	20,309,007	60	MARYLAND	8,415,165
4	TORONTO	19,816,389	61	KANSAS	8,333,205
5	COLUMBIA	19,159,897	62	CALIFORNIA, SAN DIEGO	8,324,922
6	PENNSYLVANIA STATE	17,587,527	63	CONNECTICUT	8,200,363
7	CALIFORNIA, BERKELEY	15,532,371	64	BROWN	8,064,841
8	TEXAS A&M	14,469,892	65	BOSTON	7,927,966
9	NEW YORK	14,454,550	66	LAVAL	7,803,045
10	PRINCETON	14,230,665	67	BOSTON COLLEGE	7,783,477
11	CORNELL	13,774,489	68	ILLINOIS, CHICAGO	7,729,836
12	CHICAGO	13,489,603	69	HOUSTON	7,681,826
13	EMORY	13,370,261	70	WAYNE STATE	7,665,646
14	TEXAS	13,149,394	71	YORK	7,586,224
15	PENNSYLVANIA	13,100,485	72	UTAH	7,481,035
16	PITTSBURGH	13,070,488	73	SUNY-BUFFALO	7,391,091
17	ILLINOIS, URBANA	13,053,827	74	DELAWARE	7,364,975
18	NORTH CAROLINA	12,766,914	75	TEMPLE	7,295,589
19	ALBERTA	12,689,026	76	DARTMOUTH	7,238,345
20	INDIANA	12,674,170	77	MIT	7,140,052
21	MINNESOTA	12,559,689	78	SOUTHERN ILLINOIS	7,073,858
22	WASHINGTON	12,432,102	79	LOUISIANA STATE	6,932,439
23	JOHNS HOPKINS	12,279,738	80	CALIFORNIA, IRVINE	6,926,009
24	DUKE	12,279,569	81	FLORIDA STATE	6,902,381
25	CALIFORNIA, LOS ANGELES	11,892,506	82	CASE WESTERN RESERVE	6,784,251
26	IOWA	11,866,373	83	QUEEN'S	6,767,889
27	SOUTHERN CALIFORNIA	11,857,390	84	OKLAHOMA STATE	6,690,031
28	BRITISH COLUMBIA	11,794,175	85	NEBRASKA	6,572,861
29	NORTHWESTERN	11,536,325	86	TULANE	6,525,782
30	MCGILL	11,376,632	87	ROCHESTER	6,421,998
31	ARIZONA	11,289,602	88	CALIFORNIA, DAVIS	6,417,684
32	OKLAHOMA	11,268,898	89	SASKATCHEWAN	6,382,518
33	GEORGIA	11,013,996	90	SOUTH CAROLINA	6,364,082
34	WISCONSIN	10,938,665	91	MANITOBA	6,363,752
35	FLORIDA	10,698,308	92	ALABAMA	6,274,306
36	MIAMI	10,643,708	93	COLORADO STATE	6,237,544
37	BRIGHAM YOUNG	10,301,595	94	SUNY-STONY BROOK	6,217,224
38	TENNESSEE	10,146,013	95	WATERLOO	6,194,668
39	VIRGINIA	10,082,602	96	MCMASTER	5,984,915
40	CINCINNATI	10,074,470	97	VIRGINIA TECH	5,927,765
41	RUTGERS	10,027,436	98	HAWAII	5,910,365
42	WASHINGTON U.-ST. LOUIS	9,969,715	99	MISSOURI	5,833,439
43	VANDERBILT	9,967,495	100	NEW MEXICO	5,817,037
44	ARIZONA STATE	9,875,520	101	WASHINGTON STATE	5,761,048
45	PURDUE	9,542,018	102	CALIFORNIA, SANTA BARBARA	5,712,246
46	MONTREAL	9,517,765	103	CALIFORNIA, RIVERSIDE	5,444,783
47	MICHIGAN STATE	9,487,402	104	OHIO	5,391,032
48	KENTUCKY	9,413,427	105	SYRACUSE	5,351,232
49	GEORGETOWN	9,371,271	106	OREGON	5,334,251
50	TEXAS TECH	9,269,276	107	GEORGIA TECH	5,294,007
51	GEORGE WASHINGTON	9,172,967	108	MASSACHUSETTS	5,278,206
52	OHIO STATE	8,880,060	109	AUBURN	5,261,396
53	RICE	8,878,432	110	SUNY-ALBANY	5,122,321
54	NORTH CAROLINA STATE	8,873,695	111	GUELPH	4,721,391
55	WESTERN ONTARIO	8,869,507	112	HOWARD	3,885,992
56	COLORADO	8,801,962	113	KENT STATE	3,432,288
57	NOTRE DAME	8,662,668			

Rank Order Table 7
Total Salaries & Wages Expenditures

Rank	Institution	Value
1	HARVARD	57,609,065
2	YALE	29,216,790
3	TORONTO	26,781,448
4	CALIFORNIA, BERKELEY	26,375,259
5	CALIFORNIA, LOS ANGELES	25,482,855
6	COLUMBIA	23,525,722
7	PENNSYLVANIA STATE	22,354,612
8	WISCONSIN	21,611,368
9	CORNELL	20,366,508
10	TEXAS	19,873,055
11	ILLINOIS, URBANA	18,301,128
12	RUTGERS	18,216,492
13	WASHINGTON	17,511,544
14	NEW YORK	17,353,192
15	VIRGINIA	16,911,668
16	PRINCETON	16,900,748
17	MINNESOTA	16,063,735
18	SOUTHERN CALIFORNIA	15,642,681
19	NORTH CAROLINA	15,514,366
20	PENNSYLVANIA	14,704,217
21	DUKE	14,045,905
22	OHIO STATE	14,033,069
23	INDIANA	13,915,810
24	MONTREAL	13,798,567
25	CALIFORNIA, SAN DIEGO	13,489,039
26	CONNECTICUT	12,835,917
27	BRITISH COLUMBIA	12,834,654
28	ALBERTA	12,731,858
29	FLORIDA	12,587,419
30	NORTHWESTERN	12,114,641
31	EMORY	11,941,047
32	ARIZONA	11,663,704
33	ARIZONA STATE	11,641,406
34	JOHNS HOPKINS	11,499,468
35	WASHINGTON U.-ST. LOUIS	11,287,392
36	PITTSBURGH	11,245,436
37	CHICAGO	11,179,880
38	UTAH	11,151,332
39	MIT	11,008,424
40	BRIGHAM YOUNG	10,958,814
41	IOWA	10,927,860
42	NEW MEXICO	10,920,863
43	TEXAS A&M	10,823,075
44	MARYLAND	10,614,753
45	GEORGETOWN	10,523,302
46	MICHIGAN STATE	10,483,763
47	NORTH CAROLINA STATE	9,863,503
48	SUNY-BUFFALO	9,821,025
49	MCGILL	9,714,118
50	CINCINNATI	9,670,214
51	TENNESSEE	9,480,646
52	CALIFORNIA, DAVIS	9,450,357
53	PURDUE	9,378,053
54	GEORGIA	9,158,674
55	VANDERBILT	9,028,517
56	WAYNE STATE	8,992,011
57	GEORGE WASHINGTON	8,918,119
58	NOTRE DAME	8,669,647
59	KANSAS	8,626,205
60	CALIFORNIA, IRVINE	8,520,570
61	YORK	8,471,075
62	BROWN	8,439,488
63	BOSTON	8,350,560
64	TEXAS TECH	8,343,114
65	MANITOBA	8,221,900
66	KENTUCKY	8,129,784
67	ROCHESTER	8,084,938
68	ILLINOIS, CHICAGO	8,079,971
69	LAVAL	8,054,461
70	CALIFORNIA, SANTA BARBARA	8,025,615
71	BOSTON COLLEGE	7,944,230
72	COLORADO	7,940,902
73	MIAMI	7,519,879
74	SOUTH CAROLINA	7,452,012
75	DARTMOUTH	7,437,350
76	SYRACUSE	7,423,733
77	HAWAII	7,261,998
78	DELAWARE	7,117,411
79	OREGON	7,046,392
80	IOWA STATE	6,796,494
81	MASSACHUSETTS	6,665,937
82	TEMPLE	6,563,793
83	WESTERN ONTARIO	6,445,707
84	SOUTHERN ILLINOIS	6,358,841
85	SUNY-ALBANY	6,189,138
86	SUNY-STONY BROOK	6,172,669
87	NEBRASKA	6,134,461
88	MISSOURI	6,035,326
89	WASHINGTON STATE	5,980,177
90	LOUISVILLE	5,820,338
91	HOUSTON	5,779,490
92	OHIO	5,756,808
93	QUEEN'S	5,683,656
94	CASE WESTERN RESERVE	5,611,685
95	OKLAHOMA	5,607,505
96	CALIFORNIA, RIVERSIDE	5,516,421
97	ALABAMA	5,514,096
98	LOUISIANA STATE	5,506,296
99	TULANE	5,490,584
100	FLORIDA STATE	5,465,262
101	WATERLOO	5,407,968
102	COLORADO STATE	5,391,036
103	HOWARD	5,293,792
104	SASKATCHEWAN	5,239,279
105	MCMASTER	5,200,469
106	VIRGINIA TECH	4,976,349
107	OKLAHOMA STATE	4,919,159
108	RICE	4,855,287
109	GUELPH	4,792,509
110	AUBURN	4,583,327
111	GEORGIA TECH	4,414,687
112	KENT STATE	4,413,334
113	MICHIGAN	UA

UA - Unavailable

Rank Order Table 8
Other Operating Expenditures

Rank	Institution	Value	Rank	Institution	Value
1	HARVARD	16,656,008	58	GEORGIA	2,189,915
2	WASHINGTON U.-ST. LOUIS	15,118,293	59	NEBRASKA	2,150,177
3	CALIFORNIA, BERKELEY	8,581,987	60	CINCINNATI	2,128,509
4	CALIFORNIA, LOS ANGELES	8,127,524	61	WASHINGTON STATE	2,120,059
5	YALE	7,742,895	62	TENNESSEE	2,115,532
6	CORNELL	7,216,615	63	CASE WESTERN RESERVE	2,107,843
7	COLUMBIA	7,106,593	64	OKLAHOMA STATE	2,081,186
8	WISCONSIN	6,645,300	65	KENTUCKY	2,074,958
9	TORONTO	6,144,620	66	BOSTON	2,068,658
10	MINNESOTA	5,907,137	67	LOUISVILLE	2,050,019
11	VIRGINIA	5,861,763	68	CALIFORNIA, IRVINE	2,047,034
12	SOUTHERN CALIFORNIA	5,859,726	69	GEORGE WASHINGTON	1,975,812
13	PENNSYLVANIA STATE	5,806,273	70	DARTMOUTH	1,941,533
14	NEW YORK	5,746,250	71	ROCHESTER	1,881,598
15	TEXAS	5,737,925	72	RUTGERS	1,867,124
16	WAYNE STATE	5,525,405	73	COLORADO STATE	1,856,857
17	WASHINGTON	5,519,261	74	NOTRE DAME	1,827,568
18	JOHNS HOPKINS	4,685,790	75	QUEEN'S	1,807,127
19	CALIFORNIA, SAN DIEGO	4,673,262	76	BROWN	1,748,566
20	NORTH CAROLINA STATE	4,627,288	77	ALABAMA	1,719,751
21	EMORY	4,498,938	78	COLORADO	1,710,577
22	SOUTH CAROLINA	4,453,321	79	CONNECTICUT	1,699,787
23	ARIZONA	4,290,717	80	IOWA	1,676,573
24	UTAH	4,029,797	81	CALIFORNIA, DAVIS	1,672,602
25	PRINCETON	4,001,682	82	GUELPH	1,530,113
26	PENNSYLVANIA	3,997,747	83	VIRGINIA TECH	1,527,510
27	NORTH CAROLINA	3,942,079	84	KENT STATE	1,504,737
28	YORK	3,824,183	85	OKLAHOMA	1,500,869
29	INDIANA	3,817,639	86	RICE	1,432,729
30	OHIO STATE	3,789,213	87	SUNY-BUFFALO	1,430,453
31	ILLINOIS, URBANA	3,776,614	88	CALIFORNIA, RIVERSIDE	1,385,340
32	MCGILL	3,719,813	89	MASSACHUSETTS	1,374,144
33	CALIFORNIA, SANTA BARBARA	3,706,602	90	DELAWARE	1,329,026
34	BRIGHAM YOUNG	3,507,022	91	TEMPLE	1,321,479
35	DUKE	3,369,857	92	MANITOBA	1,317,604
36	PITTSBURGH	3,297,410	93	ALBERTA	1,254,525
37	TEXAS A&M	3,193,909	94	HAWAII	1,213,637
38	TEXAS TECH	3,020,954	95	LOUISIANA STATE	1,205,455
39	FLORIDA STATE	2,947,469	96	ILLINOIS, CHICAGO	1,196,888
40	ARIZONA STATE	2,891,586	97	MONTREAL	1,169,143
41	CHICAGO	2,745,954	98	SYRACUSE	1,158,783
42	HOUSTON	2,745,186	99	IOWA STATE	1,147,520
43	OHIO	2,716,296	100	GEORGIA TECH	1,145,210
44	MARYLAND	2,681,756	101	TULANE	1,121,809
45	FLORIDA	2,631,993	102	SOUTHERN ILLINOIS	1,015,932
46	MICHIGAN STATE	2,614,114	103	BOSTON COLLEGE	1,011,046
47	NORTHWESTERN	2,599,170	104	OREGON	894,208
48	GEORGETOWN	2,593,397	105	SUNY-ALBANY	854,575
49	PURDUE	2,512,384	106	SASKATCHEWAN	828,024
50	BRITISH COLUMBIA	2,436,053	107	HOWARD	775,653
51	AUBURN	2,330,330	108	MCMASTER	714,869
52	NEW MEXICO	2,315,100	109	WATERLOO	632,989
53	MIT	2,304,151	110	WESTERN ONTARIO	604,219
54	VANDERBILT	2,276,419	111	SUNY-STONY BROOK	586,568
55	MIAMI	2,247,638	112	LAVAL	482,730
56	MISSOURI	2,239,546	113	MICHIGAN	UA
57	KANSAS	2,219,190			

UA - Unavailable

RANK ORDER TABLE 9
TOTAL LIBRARY EXPENDITURES

RANK	INSTITUTION	VALUE	RANK	INSTITUTION	VALUE
1	HARVARD	102,895,429	58	YORK	20,025,962
2	YALE	68,789,335	59	KENTUCKY	19,807,985
3	TORONTO	53,099,301	60	KANSAS	19,311,819
4	CALIFORNIA, BERKELEY	51,392,142	61	NOTRE DAME	19,249,304
5	COLUMBIA	50,451,496	62	NEW MEXICO	19,193,337
6	MICHIGAN	47,113,239	63	SUNY-BUFFALO	18,760,076
7	PENNSYLVANIA STATE	46,249,051	64	OKLAHOMA	18,569,766
8	CALIFORNIA, LOS ANGELES	46,012,476	65	COLORADO	18,549,099
9	CORNELL	41,541,659	66	BOSTON	18,533,846
10	WISCONSIN	39,482,889	67	BROWN	18,478,976
11	TEXAS	38,942,773	68	SOUTH CAROLINA	18,349,790
12	NEW YORK	38,184,768	69	CALIFORNIA, DAVIS	17,694,123
13	WASHINGTON U.-ST. LOUIS	36,484,185	70	CALIFORNIA, SANTA BARBARA	17,628,483
14	WASHINGTON	35,761,401	71	CALIFORNIA, IRVINE	17,577,307
15	PRINCETON	35,713,585	72	ILLINOIS, CHICAGO	17,080,963
16	ILLINOIS, URBANA	35,440,902	73	BOSTON COLLEGE	16,880,163
17	MINNESOTA	34,866,483	74	DARTMOUTH	16,735,903
18	SOUTHERN CALIFORNIA	33,494,903	75	IOWA STATE	16,697,188
19	VIRGINIA	32,994,275	76	LOUISVILLE	16,612,024
20	NORTH CAROLINA	32,489,076	77	ROCHESTER	16,504,838
21	PENNSYLVANIA	32,150,410	78	LAVAL	16,376,024
22	INDIANA	30,725,103	79	HOUSTON	16,319,001
23	RUTGERS	30,255,527	80	WESTERN ONTARIO	16,040,986
24	EMORY	29,951,903	81	MANITOBA	15,996,061
25	DUKE	29,865,742	82	DELAWARE	15,893,753
26	TEXAS A&M	28,646,821	83	FLORIDA STATE	15,385,125
27	JOHNS HOPKINS	28,565,767	84	TEMPLE	15,273,023
28	PITTSBURGH	27,884,821	85	RICE	15,213,029
29	CHICAGO	27,743,553	86	NEBRASKA	15,016,679
30	ARIZONA	27,342,719	87	CASE WESTERN RESERVE	14,597,197
31	BRITISH COLUMBIA	27,218,279	88	SOUTHERN ILLINOIS	14,589,368
32	OHIO STATE	26,954,787	89	HAWAII	14,586,000
33	CALIFORNIA, SAN DIEGO	26,794,963	90	QUEEN'S	14,352,701
34	ALBERTA	26,769,667	91	MISSOURI	14,252,068
35	NORTHWESTERN	26,491,033	92	SYRACUSE	13,995,085
36	FLORIDA	26,216,130	93	WASHINGTON STATE	13,934,946
37	MCGILL	24,980,527	94	OHIO	13,918,324
38	BRIGHAM YOUNG	24,971,681	95	OKLAHOMA STATE	13,760,638
39	MONTREAL	24,702,567	96	LOUISIANA STATE	13,712,738
40	IOWA	24,648,673	97	COLORADO STATE	13,602,053
41	ARIZONA STATE	24,581,554	98	ALABAMA	13,587,699
42	NORTH CAROLINA STATE	23,548,957	99	MASSACHUSETTS	13,415,087
43	CONNECTICUT	22,980,420	100	OREGON	13,371,795
44	UTAH	22,833,522	101	TULANE	13,262,398
45	MICHIGAN STATE	22,811,257	102	SUNY-STONY BROOK	13,030,856
46	GEORGIA	22,679,865	103	CALIFORNIA, RIVERSIDE	12,590,290
47	GEORGETOWN	22,614,466	104	SASKATCHEWAN	12,524,891
48	WAYNE STATE	22,261,534	105	VIRGINIA TECH	12,508,979
49	CINCINNATI	21,992,639	106	WATERLOO	12,290,479
50	TENNESSEE	21,871,331	107	AUBURN	12,251,864
51	MARYLAND	21,841,492	108	SUNY-ALBANY	12,248,462
52	PURDUE	21,530,300	109	MCMASTER	12,052,587
53	VANDERBILT	21,396,174	110	GUELPH	11,069,118
54	TEXAS TECH	20,733,480	111	GEORGIA TECH	10,906,711
55	MIT	20,663,012	112	HOWARD	10,012,937
56	MIAMI	20,562,774	113	KENT STATE	9,436,048
57	GEORGE WASHINGTON	20,210,171			

RANK ORDER TABLE 10
MONOGRAPHS PURCHASED (VOLUMES)

RANK	INSTITUTION	VALUE	RANK	INSTITUTION	VALUE
1	COLUMBIA	249,680	58	CALIFORNIA, IRVINE	29,659
2	YALE	207,276	59	IOWA STATE	29,029
3	TORONTO	142,023	60	DELAWARE	28,892
4	CHICAGO	98,811	61	TENNESSEE	28,860
5	PRINCETON	88,070	62	CALIFORNIA, SANTA BARBARA	28,698
6	ILLINOIS, URBANA	77,420	63	WESTERN ONTARIO	28,444
7	TEXAS A&M	73,484	64	TEMPLE	28,418
8	LOUISIANA STATE	70,185	65	OKLAHOMA	28,067
9	OHIO STATE	67,494	66	SOUTH CAROLINA	28,003
10	CALIFORNIA, LOS ANGELES	65,953	67	WASHINGTON U.-ST. LOUIS	27,942
11	IOWA	63,627	68	MISSOURI	27,314
12	INDIANA	63,156	69	DARTMOUTH	26,948
13	ARIZONA STATE	60,160	70	VANDERBILT	26,592
14	DUKE	59,547	71	CONNECTICUT	26,317
15	PENNSYLVANIA STATE	59,347	72	NORTH CAROLINA STATE	24,660
16	MARYLAND	59,002	73	OKLAHOMA STATE	23,897
17	NORTH CAROLINA	57,520	74	NEW MEXICO	23,806
18	MONTREAL	57,485	75	AUBURN	23,579
19	NEW YORK	56,881	76	RUTGERS	23,331
20	WASHINGTON	55,430	77	GEORGE WASHINGTON	22,076
21	WISCONSIN	54,891	78	PURDUE	21,860
22	ALABAMA	50,909	79	QUEEN'S	21,149
23	TEXAS	49,716	80	SASKATCHEWAN	21,137
24	ARIZONA	49,672	81	OREGON	20,782
25	FLORIDA	47,038	82	ROCHESTER	20,753
26	GEORGIA	46,157	83	HAWAII	20,707
27	MCMASTER	46,059	84	VIRGINIA TECH	20,441
28	KENTUCKY	46,041	85	SYRACUSE	19,562
29	BRIGHAM YOUNG	43,681	86	WAYNE STATE	19,277
30	CALIFORNIA, SAN DIEGO	43,233	87	MIT	19,207
31	OHIO	42,950	88	FLORIDA STATE	18,726
32	CINCINNATI	42,385	89	WASHINGTON STATE	18,694
33	MANITOBA	41,616	90	WATERLOO	17,868
34	MCGILL	40,229	91	BOSTON	17,181
35	EMORY	40,009	92	SOUTHERN ILLINOIS	16,532
36	NORTHWESTERN	39,181	93	NEBRASKA	16,361
37	VIRGINIA	38,893	94	SUNY-BUFFALO	15,436
38	TEXAS TECH	38,588	95	GUELPH	14,261
39	KANSAS	38,123	96	SUNY-ALBANY	14,039
40	COLORADO	35,690	97	CASE WESTERN RESERVE	13,840
41	LOUISVILLE	35,345	98	SUNY-STONY BROOK	12,602
42	CALIFORNIA, RIVERSIDE	35,287	99	KENT STATE	10,752
43	MIAMI	35,119	100	GEORGIA TECH	5,827
44	GEORGETOWN	34,936	101	HOWARD	3,689
45	YORK	34,747	102	ALBERTA	UA
46	RICE	34,629	103	BOSTON COLLEGE	UA
47	MICHIGAN STATE	34,427	104	BRITISH COLUMBIA	UA
48	SOUTHERN CALIFORNIA	34,422	105	CALIFORNIA, BERKELEY	UA
49	MINNESOTA	34,075	106	CALIFORNIA, DAVIS	UA
50	BROWN	33,405	107	CORNELL	UA
51	ILLINOIS, CHICAGO	32,712	108	HARVARD	UA
52	LAVAL	32,600	109	MASSACHUSETTS	UA
53	TULANE	31,967	110	MICHIGAN	UA
54	HOUSTON	31,384	111	PENNSYLVANIA	UA
55	JOHNS HOPKINS	30,874	112	PITTSBURGH	UA
56	NOTRE DAME	30,274	113	UTAH	UA
57	COLORADO STATE	30,217			

UA – Unavailable

RANK ORDER TABLE 11
EXPENDITURES FOR MONOGRAPHS

RANK	INSTITUTION	VALUE	RANK	INSTITUTION	VALUE
1	YALE	11,104,116	58	WESTERN ONTARIO	1,767,990
2	HARVARD	9,595,629	59	NORTH CAROLINA STATE	1,732,653
3	CALIFORNIA, BERKELEY	7,385,144	60	ALABAMA	1,712,729
4	TORONTO	6,929,869	61	DARTMOUTH	1,712,646
5	COLUMBIA	6,035,280	62	SASKATCHEWAN	1,697,999
6	PRINCETON	5,693,652	63	MARYLAND	1,665,420
7	CORNELL	5,234,477	64	VANDERBILT	1,658,666
8	TEXAS A&M	5,202,466	65	GEORGETOWN	1,627,048
9	CALIFORNIA, LOS ANGELES	4,620,063	66	BOSTON COLLEGE	1,618,129
10	CHICAGO	4,360,625	67	COLORADO STATE	1,587,712
11	ALBERTA	4,295,457	68	COLORADO	1,584,992
12	EMORY	4,172,594	69	CALIFORNIA, RIVERSIDE	1,558,422
13	INDIANA	3,966,347	70	LAVAL	1,546,700
14	TEXAS	3,684,604	71	JOHNS HOPKINS	1,545,873
15	PENNSYLVANIA	3,320,341	72	TULANE	1,536,290
16	RICE	3,308,947	73	KENTUCKY	1,532,372
17	OKLAHOMA	3,291,713	74	PURDUE	1,498,189
18	PENNSYLVANIA STATE	3,230,617	75	OKLAHOMA STATE	1,410,759
19	NEW YORK	3,169,855	76	MANITOBA	1,405,048
20	WASHINGTON	3,078,929	77	WASHINGTON U.-ST. LOUIS	1,397,670
21	NORTHWESTERN	3,078,747	78	RUTGERS	1,351,966
22	PITTSBURGH	3,063,051	79	WAYNE STATE	1,349,418
23	DUKE	3,020,512	80	OREGON	1,335,830
24	GEORGE WASHINGTON	3,007,868	81	HOUSTON	1,331,054
25	NOTRE DAME	2,941,063	82	FLORIDA STATE	1,319,330
26	CINCINNATI	2,901,075	83	LOUISVILLE	1,312,102
27	NORTH CAROLINA	2,786,918	84	MIT	1,282,263
28	MCGILL	2,756,401	85	HAWAII	1,243,398
29	ILLINOIS, URBANA	2,644,792	86	VIRGINIA TECH	1,241,043
30	CALIFORNIA, SAN DIEGO	2,640,757	87	SOUTH CAROLINA	1,236,288
31	WISCONSIN	2,532,640	88	MCMASTER	1,218,069
32	ARIZONA	2,495,290	89	TEMPLE	1,206,321
33	IOWA	2,415,944	90	QUEEN'S	1,120,089
34	MINNESOTA	2,411,411	91	OHIO	1,111,982
35	MONTREAL	2,394,073	92	ILLINOIS, CHICAGO	1,097,522
36	YORK	2,350,167	93	CONNECTICUT	1,091,820
37	BROWN	2,335,426	94	SUNY-BUFFALO	1,043,075
38	OHIO STATE	2,288,345	95	GUELPH	1,036,737
39	LOUISIANA STATE	2,258,247	96	ROCHESTER	1,011,000
40	MIAMI	2,242,161	97	SYRACUSE	988,995
41	FLORIDA	2,232,834	98	MISSOURI	964,293
42	TEXAS TECH	2,213,875	99	SOUTHERN ILLINOIS	962,335
43	GEORGIA	2,151,225	100	MASSACHUSETTS	910,277
44	ARIZONA STATE	2,150,937	101	NEBRASKA	900,763
45	DELAWARE	2,149,791	102	WASHINGTON STATE	897,590
46	VIRGINIA	2,128,109	103	AUBURN	891,275
47	BRIGHAM YOUNG	2,106,921	104	SUNY-ALBANY	888,311
48	CALIFORNIA, SANTA BARBARA	2,076,688	105	BOSTON	887,337
49	MICHIGAN STATE	2,053,447	106	WATERLOO	878,710
50	CALIFORNIA, DAVIS	2,033,057	107	NEW MEXICO	850,138
51	SOUTHERN CALIFORNIA	2,024,772	108	CASE WESTERN RESERVE	772,147
52	BRITISH COLUMBIA	1,969,804	109	SUNY-STONY BROOK	629,154
53	TENNESSEE	1,907,902	110	KENT STATE	622,190
54	UTAH	1,858,489	111	GEORGIA TECH	490,380
55	KANSAS	1,811,413	112	HOWARD	239,204
56	IOWA STATE	1,794,752	113	MICHIGAN	UA
57	CALIFORNIA, IRVINE	1,784,841			

UA – Unavailable

Rank Order Table 12
Current Serials Purchased (Subscriptions)

Rank	Institution	Value	Rank	Institution	Value
1	FLORIDA	68,194	58	CALIF., SAN DIEGO	20,543
2	YALE	68,116	59	OHIO STATE	20,212
3	CORNELL	65,923	60	KENTUCKY	20,073
4	INDIANA	65,593	61	MANITOBA	20,005
5	PRINCETON	62,921	62	COLORADO STATE	19,993
6	TORONTO	61,357	63	HAWAII	19,789
7	COLUMBIA	59,936	64	CALIF., RIVERSIDE	19,739
8	PENNSYLVANIA STATE	54,155	65	SASKATCHEWAN	19,632
9	RUTGERS	47,060	66	TEMPLE	18,717
10	TEXAS A&M	45,079	67	HOUSTON	18,483
11	WISCONSIN	44,710	68	BRIGHAM YOUNG	18,426
12	JOHNS HOPKINS	43,014	69	SYRACUSE	16,862
13	DARTMOUTH	42,880	70	IOWA STATE	16,698
14	MCGILL	42,794	71	OHIO	16,346
15	WESTERN ONTARIO	41,912	72	GEORGE WASHINGTON	15,855
16	PENNSYLVANIA	41,166	73	TULANE	15,710
17	NEW YORK	41,121	74	MCMASTER	14,921
18	FLORIDA STATE	40,475	75	PURDUE	13,868
19	SUNY-ALBANY	38,859	76	OREGON	13,033
20	MONTREAL	38,444	77	MARYLAND	12,847
21	BOSTON COLLEGE	37,171	78	MIT	12,570
22	CINCINNATI	36,311	79	GEORGIA TECH	12,506
23	NEBRASKA	35,714	80	CASE WESTERN RESERVE	12,235
24	MASSACHUSETTS	34,199	81	NEW MEXICO	11,891
25	MISSOURI	32,902	82	GUELPH	11,317
26	DUKE	31,892	83	KENT STATE	11,176
27	OKLAHOMA	31,752	84	RICE	11,010
28	IOWA	31,605	85	AUBURN	10,792
29	VIRGINIA TECH	31,365	86	COLORADO	10,656
30	MINNESOTA	31,098	87	HOWARD	9,764
31	CONNECTICUT	30,295	88	DELAWARE	7,518
32	SOUTHERN CALIFORNIA	29,916	89	ARIZONA	UA
33	GEORGIA	29,047	90	BOSTON	UA
34	NORTH CAROLINA	28,862	91	BRITISH COLUMBIA	UA
35	VIRGINIA	28,375	92	CALIF., BERKELEY	UA
36	ALBERTA	28,257	93	CALIF., DAVIS	UA
37	KANSAS	27,721	94	CALIF., LOS ANGELES	UA
38	SUNY-STONY BROOK	27,651	95	CHICAGO	UA
39	TEXAS	26,433	96	EMORY	UA
40	ILLINOIS, CHICAGO	26,350	97	HARVARD	UA
41	TENNESSEE	25,650	98	ILLINOIS, URBANA	UA
42	VANDERBILT	25,476	99	LOUISIANA STATE	UA
43	SOUTHERN ILLINOIS	25,420	100	MIAMI	UA
44	TEXAS TECH	25,193	101	MICHIGAN	UA
45	GEORGETOWN	24,115	102	NORTHWESTERN	UA
46	YORK	23,686	103	NOTRE DAME	UA
47	NORTH CAROLINA STATE	23,011	104	OKLAHOMA STATE	UA
48	LOUISVILLE	22,833	105	PITTSBURGH	UA
49	ALABAMA	22,663	106	ROCHESTER	UA
50	SUNY-BUFFALO	22,145	107	SOUTH CAROLINA	UA
51	CALIF., SANTA BARBARA	22,084	108	UTAH	UA
52	BROWN	21,888	109	WASHINGTON	UA
53	LAVAL	21,438	110	WASHINGTON STATE	UA
54	MICHIGAN STATE	21,326	111	WASHINGTON U.-ST. LOUIS	UA
55	ARIZONA STATE	21,282	112	WATERLOO	UA
56	QUEEN'S	20,947	113	WAYNE STATE	UA
57	CALIF., IRVINE	20,649			

UA – Unavailable

Rank Order Table 13
Expenditures for Current Serials

Rank	Institution	Value	Rank	Institution	Value
1	COLUMBIA	11,360,265	58	SUNY-BUFFALO	5,873,962
2	HARVARD	10,257,953	59	BOSTON COLLEGE	5,824,029
3	JOHNS HOPKINS	9,881,835	60	MIT	5,802,131
4	NEW YORK	9,715,687	61	HOUSTON	5,670,011
5	ILLINOIS, URBANA	9,520,111	62	KANSAS	5,634,506
6	TORONTO	9,446,724	63	SOUTHERN ILLINOIS	5,488,465
7	MINNESOTA	9,262,586	64	DARTMOUTH	5,466,183
8	PENNSYLVANIA	9,222,287	65	CASE WESTERN RESERVE	5,443,602
9	WASHINGTON	9,086,013	66	UTAH	5,422,996
10	CHICAGO	8,897,582	67	EMORY	5,405,397
11	TEXAS A&M	8,678,296	68	OKLAHOMA	5,375,805
12	BRITISH COLUMBIA	8,431,982	69	CALIFORNIA, SAN DIEGO	5,369,676
13	DUKE	8,286,199	70	NEBRASKA	5,180,965
14	PITTSBURGH	8,182,638	71	QUEEN'S	5,141,295
15	VANDERBILT	8,088,649	72	GEORGE WASHINGTON	5,102,310
16	BRIGHAM YOUNG	8,011,243	73	BROWN	5,040,773
17	CALIFORNIA, BERKELEY	7,947,095	74	ROCHESTER	5,024,194
18	TENNESSEE	7,940,141	75	NOTRE DAME	5,012,370
19	YALE	7,891,785	76	SOUTH CAROLINA	4,941,208
20	FLORIDA	7,867,232	77	DELAWARE	4,934,128
21	TEXAS	7,862,167	78	YORK	4,835,964
22	ALBERTA	7,837,627	79	RICE	4,767,337
23	ARIZONA	7,743,061	80	SUNY-STONY BROOK	4,757,927
24	NORTH CAROLINA	7,608,439	81	CALIFORNIA, IRVINE	4,757,232
25	NORTHWESTERN	7,537,298	82	OHIO STATE	4,738,482
26	CORNELL	7,500,212	83	TULANE	4,662,815
27	GEORGIA	7,465,945	84	MISSOURI	4,636,091
28	RUTGERS	7,364,328	85	WASHINGTON STATE	4,628,803
29	WASHINGTON U.-ST. LOUIS	7,203,389	86	HAWAII	4,573,379
30	WESTERN ONTARIO	7,101,517	87	TEMPLE	4,545,975
31	MICHIGAN STATE	7,090,194	88	SASKATCHEWAN	4,528,334
32	LOUISVILLE	7,078,692	89	ALABAMA	4,515,297
33	IOWA	7,017,808	90	NEW MEXICO	4,447,558
34	MIAMI	6,944,650	91	LOUISIANA STATE	4,386,085
35	SOUTHERN CALIFORNIA	6,941,777	92	FLORIDA STATE	4,336,516
36	MONTREAL	6,862,460	93	MCMASTER	4,312,240
37	KENTUCKY	6,854,003	94	COLORADO STATE	4,279,800
38	PRINCETON	6,782,562	95	VIRGINIA TECH	4,253,860
39	ARIZONA STATE	6,758,072	96	MANITOBA	4,240,389
40	GEORGETOWN	6,743,739	97	SYRACUSE	4,136,091
41	PURDUE	6,687,528	98	WATERLOO	4,129,732
42	CALIFORNIA, LOS ANGELES	6,685,201	99	MASSACHUSETTS	4,094,358
43	CONNECTICUT	6,564,118	100	OKLAHOMA STATE	4,071,623
44	MCGILL	6,518,215	101	MARYLAND	3,996,874
45	VIRGINIA	6,464,325	102	SUNY-ALBANY	3,886,662
46	COLORADO	6,398,185	103	CALIFORNIA, RIVERSIDE	3,799,154
47	IOWA STATE	6,337,530	104	OHIO	3,788,272
48	TEXAS TECH	6,324,203	105	CALIFORNIA, SANTA BARBARA	3,635,558
49	BOSTON	6,184,327	106	GEORGIA TECH	3,614,169
50	CINCINNATI	6,178,530	107	HOWARD	3,501,658
51	INDIANA	6,147,597	108	CALIFORNIA, DAVIS	3,461,034
52	ILLINOIS, CHICAGO	6,087,267	109	AUBURN	3,440,754
53	WISCONSIN	6,060,813	110	OREGON	3,367,229
54	LAVAL	5,994,235	111	GUELPH	3,047,922
55	WAYNE STATE	5,991,513	112	KENT STATE	2,738,556
56	PENNSYLVANIA STATE	5,933,378	113	MICHIGAN	UA
57	NORTH CAROLINA STATE	5,875,549			

UA - Unavailable

RANK ORDER TABLE 14
TOTAL ITEMS LOANED (ILL/DD)

RANK	INSTITUTION	VALUE	RANK	INSTITUTION	VALUE
1	MINNESOTA	180,504	58	IOWA STATE	34,603
2	OHIO STATE	145,272	59	BRIGHAM YOUNG	34,391
3	WISCONSIN	109,100	60	DARTMOUTH	34,139
4	ALBERTA	97,042	61	MIAMI	33,755
5	ILLINOIS, URBANA	90,304	62	CALIFORNIA, DAVIS	33,112
6	OHIO	88,448	63	NORTHWESTERN	32,515
7	INDIANA	86,289	64	TEXAS TECH	31,270
8	WASHINGTON	79,369	65	OKLAHOMA	30,669
9	OREGON	75,939	66	HOUSTON	29,435
10	ILLINOIS, CHICAGO	74,438	67	CALIFORNIA, IRVINE	29,067
11	PITTSBURGH	74,175	68	PURDUE	29,018
12	WATERLOO	72,073	69	NEBRASKA	28,123
13	CALIFORNIA, LOS ANGELES	59,765	70	MONTREAL	27,800
14	CORNELL	59,431	71	VANDERBILT	27,692
15	ARIZONA STATE	59,341	72	MANITOBA	27,540
16	MISSOURI	58,619	73	ROCHESTER	26,482
17	WAYNE STATE	58,149	74	QUEEN'S	26,157
18	IOWA	58,129	75	SASKATCHEWAN	25,924
19	CINCINNATI	57,635	76	RUTGERS	25,699
20	PENNSYLVANIA STATE	57,373	77	BRITISH COLUMBIA	24,718
21	CALIFORNIA, SAN DIEGO	56,644	78	LOUISVILLE	24,666
22	NORTH CAROLINA	54,840	79	CALIFORNIA, SANTA BARBARA	24,637
23	COLUMBIA	54,523	80	TULANE	24,484
24	CONNECTICUT	53,812	81	SOUTHERN CALIFORNIA	24,066
25	SOUTHERN ILLINOIS	53,688	82	CALIFORNIA, RIVERSIDE	23,429
26	GEORGIA	51,995	83	NOTRE DAME	23,323
27	COLORADO	51,802	84	SUNY-STONY BROOK	23,160
28	KANSAS	51,672	85	TEMPLE	21,599
29	TEXAS	51,621	86	VIRGINIA TECH	21,585
30	ARIZONA	50,161	87	SOUTH CAROLINA	21,503
31	COLORADO STATE	49,178	88	NEW MEXICO	21,450
32	MICHIGAN STATE	48,702	89	FLORIDA	21,411
33	CHICAGO	46,516	90	LAVAL	21,326
34	GUELPH	46,319	91	BOSTON COLLEGE	19,973
35	CALIFORNIA, BERKELEY	46,310	92	WESTERN ONTARIO	18,307
36	TEXAS A&M	45,758	93	MCGILL	17,475
37	FLORIDA STATE	45,473	94	DELAWARE	17,323
38	HARVARD	44,721	95	BOSTON	17,254
39	YALE	43,006	96	MARYLAND	16,818
40	EMORY	42,853	97	MCMASTER	16,093
41	VIRGINIA	42,612	98	HOWARD	16,069
42	UTAH	42,595	99	SYRACUSE	15,739
43	MASSACHUSETTS	41,537	100	NEW YORK	15,236
44	PENNSYLVANIA	41,307	101	YORK	14,783
45	DUKE	41,107	102	NORTH CAROLINA STATE	13,003
46	WASHINGTON U.-ST. LOUIS	40,603	103	KENT STATE	12,825
47	CASE WESTERN RESERVE	40,160	104	RICE	12,704
48	JOHNS HOPKINS	37,960	105	MIT	12,518
49	TENNESSEE	37,938	106	AUBURN	12,456
50	SUNY-BUFFALO	37,536	107	ALABAMA	12,053
51	OKLAHOMA STATE	37,506	108	SUNY-ALBANY	10,662
52	GEORGETOWN	37,201	109	LOUISIANA STATE	9,574
53	GEORGE WASHINGTON	36,864	110	GEORGIA TECH	9,269
54	TORONTO	36,580	111	HAWAII	2,513
55	BROWN	36,325	112	MICHIGAN	UA
56	KENTUCKY	35,541	113	WASHINGTON STATE	UA
57	PRINCETON	34,790			

UA – Unavailable

RANK ORDER TABLE 15
TOTAL ITEMS BORROWED (ILL/DD)

RANK	INSTITUTION	VALUE	RANK	INSTITUTION	VALUE
1	OHIO STATE	101,722	58	FLORIDA	24,955
2	ILLINOIS, URBANA	91,596	59	CHICAGO	24,920
3	PENNSYLVANIA	72,214	60	NEW YORK	24,607
4	OHIO	71,084	61	UTAH	24,603
5	GUELPH	66,594	62	VIRGINIA	24,344
6	COLORADO STATE	65,964	63	SASKATCHEWAN	24,295
7	WISCONSIN	65,313	64	SYRACUSE	23,957
8	OREGON	61,671	65	BRITISH COLUMBIA	23,666
9	TEXAS A&M	61,658	66	CALIFORNIA, BERKELEY	22,800
10	GEORGE WASHINGTON	58,194	67	GEORGETOWN	22,604
11	PENNSYLVANIA STATE	53,146	68	PITTSBURGH	22,394
12	COLUMBIA	53,142	69	NEW MEXICO	22,114
13	WASHINGTON U.-ST. LOUIS	51,110	70	ROCHESTER	21,231
14	WASHINGTON	50,498	71	VIRGINIA TECH	21,218
15	CINCINNATI	50,115	72	SOUTH CAROLINA	21,166
16	WATERLOO	46,513	73	DUKE	21,044
17	MISSOURI	44,953	74	DELAWARE	20,933
18	INDIANA	41,826	75	FLORIDA STATE	20,733
19	CALIFORNIA, LOS ANGELES	41,077	76	SUNY-BUFFALO	20,440
20	MARYLAND	40,017	77	OKLAHOMA STATE	20,089
21	CASE WESTERN RESERVE	39,709	78	SUNY-ALBANY	20,088
22	ILLINOIS, CHICAGO	39,368	79	MONTREAL	19,926
23	BRIGHAM YOUNG	38,362	80	VANDERBILT	19,547
24	MANITOBA	38,329	81	HOUSTON	19,463
25	KANSAS	38,009	82	BOSTON	18,007
26	JOHNS HOPKINS	36,374	83	EMORY	17,634
27	NORTHWESTERN	36,275	84	MIAMI	16,918
28	CALIFORNIA, DAVIS	35,948	85	CALIFORNIA, RIVERSIDE	16,827
29	MINNESOTA	35,615	86	TEMPLE	16,491
30	OKLAHOMA	34,719	87	IOWA STATE	16,389
31	CONNECTICUT	34,709	88	NOTRE DAME	16,011
32	MASSACHUSETTS	34,507	89	LAVAL	15,520
33	BROWN	34,474	90	MCMASTER	15,370
34	ALBERTA	33,776	91	NEBRASKA	15,260
35	CALIFORNIA, SANTA BARBARA	32,804	92	NORTH CAROLINA	14,947
36	SOUTHERN ILLINOIS	32,660	93	HARVARD	14,880
37	PRINCETON	32,341	94	SUNY-STONY BROOK	14,749
38	CORNELL	32,339	95	QUEEN'S	13,349
39	MICHIGAN STATE	31,527	96	RICE	12,984
40	TEXAS TECH	30,851	97	TULANE	12,701
41	TEXAS	30,097	98	BOSTON COLLEGE	12,473
42	TENNESSEE	30,075	99	TORONTO	12,127
43	WAYNE STATE	29,966	100	LOUISVILLE	12,119
44	DARTMOUTH	29,766	101	MIT	11,443
45	KENTUCKY	29,300	102	SOUTHERN CALIFORNIA	11,351
46	ARIZONA	29,212	103	ALABAMA	9,997
47	COLORADO	29,054	104	GEORGIA	9,378
48	YALE	28,800	105	AUBURN	8,748
49	CALIFORNIA, SAN DIEGO	28,753	106	KENT STATE	8,420
50	CALIFORNIA, IRVINE	28,323	107	HAWAII	8,178
51	LOUISIANA STATE	28,199	108	WESTERN ONTARIO	5,780
52	PURDUE	27,862	109	HOWARD	5,102
53	IOWA	27,744	110	GEORGIA TECH	5,045
54	ARIZONA STATE	27,541	111	YORK	4,921
55	NORTH CAROLINA STATE	26,813	112	MICHIGAN	UA
56	RUTGERS	25,729	113	WASHINGTON STATE	UA
57	MCGILL	25,718			

UA – Unavailable

Rank Order Table 16
Professional Staff (FTE)

Rank	Institution	Value	Rank	Institution	Value
1	HARVARD	504	58	CONNECTICUT	79
2	YALE	254	59	GEORGIA	79
3	COLUMBIA	239	60	BOSTON COLLEGE	76
4	WISCONSIN	229	61	UTAH	76
5	CALIFORNIA, BERKELEY	212	62	GEORGE WASHINGTON	74
6	ILLINOIS, URBANA	186	63	MICHIGAN STATE	74
7	TORONTO	185	64	WESTERN ONTARIO	73
8	CALIFORNIA, LOS ANGELES	169	65	ILLINOIS, CHICAGO	72
9	INDIANA	163	66	HAWAII	69
10	MICHIGAN	160	67	NEW MEXICO	69
11	DUKE	156	68	FLORIDA STATE	68
12	PENNSYLVANIA STATE	149	69	SOUTH CAROLINA	68
13	TEXAS	149	70	MCGILL	66
14	PRINCETON	145	71	PURDUE	66
15	NORTH CAROLINA	141	72	SYRACUSE	65
16	WASHINGTON	138	73	CHICAGO	64
17	SUNY-BUFFALO	137	74	OHIO	64
18	JOHNS HOPKINS	135	75	CASE WESTERN RESERVE	63
19	OHIO STATE	134	76	OKLAHOMA STATE	63
20	TEXAS A&M	132	77	CALIFORNIA, DAVIS	62
21	MARYLAND	127	78	TEMPLE	62
22	SOUTHERN CALIFORNIA	125	79	CALIFORNIA, IRVINE	61
23	NORTH CAROLINA STATE	124	80	MANITOBA	61
24	CORNELL	119	81	OKLAHOMA	61
25	NEW YORK	119	82	ALABAMA	60
26	PENNSYLVANIA	119	83	LAVAL	60
27	PITTSBURGH	118	84	YORK	60
28	FLORIDA	116	85	SUNY-ALBANY	59
29	NORTHWESTERN	116	86	DARTMOUTH	58
30	BRIGHAM YOUNG	114	87	NOTRE DAME	58
31	CINCINNATI	111	88	DELAWARE	57
32	MONTREAL	110	89	RICE	57
33	WASHINGTON U.-ST. LOUIS	108	90	MISSOURI	56
34	RUTGERS	105	91	SOUTHERN ILLINOIS	56
35	KANSAS	103	92	MASSACHUSETTS	55
36	VANDERBILT	103	93	HOWARD	54
37	MINNESOTA	102	94	TULANE	54
38	VIRGINIA	102	95	OREGON	53
39	ARIZONA	100	96	NEBRASKA	52
40	TEXAS TECH	100	97	AUBURN	51
41	BRITISH COLUMBIA	99	98	COLORADO	51
42	WAYNE STATE	99	99	IOWA STATE	51
43	KENTUCKY	97	100	LOUISIANA STATE	51
44	MIT	96	101	LOUISVILLE	51
45	BOSTON	95	102	CALIFORNIA, SANTA BARBARA	49
46	EMORY	93	103	GEORGIA TECH	48
47	CALIFORNIA, SAN DIEGO	92	104	SASKATCHEWAN	47
48	MIAMI	92	105	WASHINGTON STATE	47
49	ROCHESTER	92	106	COLORADO STATE	46
50	TENNESSEE	92	107	QUEEN'S	45
51	ARIZONA STATE	90	108	CALIFORNIA, RIVERSIDE	43
52	SUNY-STONY BROOK	90	109	GUELPH	39
53	IOWA	89	110	VIRGINIA TECH	36
54	HOUSTON	87	111	WATERLOO	35
55	GEORGETOWN	83	112	MCMASTER	34
56	ALBERTA	80	113	KENT STATE	32
57	BROWN	80			

RANK ORDER TABLE 17
SUPPORT STAFF (FTE)

RANK	INSTITUTION	VALUE	RANK	INSTITUTION	VALUE
1	HARVARD	583	58	KENTUCKY	119
2	PENNSYLVANIA STATE	399	59	TEXAS TECH	119
3	YALE	370	60	COLORADO	117
4	TORONTO	351	61	DARTMOUTH	116
5	CORNELL	313	62	MIAMI	115
6	MICHIGAN	313	63	MISSOURI	115
7	TEXAS	311	64	NEBRASKA	114
8	CALIFORNIA, LOS ANGELES	282	65	YORK	112
9	MONTREAL	282	66	QUEEN'S	110
10	COLUMBIA	263	67	KANSAS	109
11	RUTGERS	240	68	NORTH CAROLINA STATE	108
12	NEW YORK	226	69	SOUTH CAROLINA	108
13	ALBERTA	214	70	VANDERBILT	108
14	MINNESOTA	214	71	DELAWARE	107
15	VIRGINIA	214	72	MCMASTER	106
16	WASHINGTON	214	73	TULANE	106
17	PRINCETON	210	74	BOSTON	102
18	ARIZONA STATE	207	75	WESTERN ONTARIO	102
19	NORTH CAROLINA	205	76	BROWN	101
20	ILLINOIS, URBANA	202	77	LOUISIANA STATE	99
21	BRITISH COLUMBIA	199	78	OREGON	99
22	FLORIDA	196	79	SASKATCHEWAN	99
23	UTAH	195	80	WATERLOO	99
24	CALIFORNIA, BERKELEY	188	81	MIT	95
25	CALIFORNIA, SAN DIEGO	185	82	IOWA STATE	93
26	GEORGIA	185	83	SOUTHERN ILLINOIS	93
27	INDIANA	180	84	OKLAHOMA	92
28	CHICAGO	178	85	LOUISVILLE	91
29	PENNSYLVANIA	177	86	TEMPLE	91
30	WISCONSIN	177	87	CALIFORNIA, RIVERSIDE	90
31	PITTSBURGH	167	88	MARYLAND	90
32	MCGILL	163	89	VIRGINIA TECH	90
33	JOHNS HOPKINS	161	90	HAWAII	89
34	NEW MEXICO	160	91	WAYNE STATE	89
35	EMORY	159	92	WASHINGTON STATE	84
36	NOTRE DAME	158	93	CONNECTICUT	82
37	ARIZONA	156	94	ALABAMA	81
38	OHIO STATE	156	95	OKLAHOMA STATE	80
39	LAVAL	155	96	BOSTON COLLEGE	77
40	DUKE	154	97	CINCINNATI	75
41	CALIFORNIA, DAVIS	150	98	HOWARD	75
42	TENNESSEE	141	99	MASSACHUSETTS	75
43	TEXAS A&M	141	100	GUELPH	74
44	IOWA	140	101	GEORGIA TECH	70
45	MANITOBA	139	102	SUNY-ALBANY	70
46	NORTHWESTERN	139	103	HOUSTON	68
47	CALIFORNIA, SANTA BARBARA	137	104	OHIO	68
48	SYRACUSE	135	105	ROCHESTER	67
49	PURDUE	133	106	RICE	63
50	WASHINGTON U.-ST. LOUIS	133	107	COLORADO STATE	60
51	GEORGETOWN	132	108	AUBURN	57
52	FLORIDA STATE	129	109	BRIGHAM YOUNG	57
53	ILLINOIS, CHICAGO	123	110	CASE WESTERN RESERVE	55
54	SOUTHERN CALIFORNIA	122	111	SUNY-BUFFALO	50
55	CALIFORNIA, IRVINE	121	112	KENT STATE	48
56	MICHIGAN STATE	121	113	SUNY-STONY BROOK	29
57	GEORGE WASHINGTON	120			

RANK ORDER TABLE 18
TOTAL STAFF (FTE)

RANK	INSTITUTION	VALUE	RANK	INSTITUTION	VALUE
1	HARVARD	1,198	58	SUNY-BUFFALO	265
2	YALE	702	59	WAYNE STATE	265
3	TORONTO	663	60	PURDUE	255
4	PENNSYLVANIA STATE	625	61	NOTRE DAME	252
5	CALIFORNIA, BERKELEY	624	62	VANDERBILT	252
6	CALIFORNIA, LOS ANGELES	604	63	GEORGE WASHINGTON	249
7	COLUMBIA	602	64	FLORIDA STATE	245
8	MICHIGAN	596	65	CALIFORNIA, IRVINE	242
9	WISCONSIN	588	66	CINCINNATI	240
10	TEXAS	572	67	YORK	233
11	CORNELL	540	68	MANITOBA	227
12	ILLINOIS, URBANA	515	69	SYRACUSE	227
13	WASHINGTON	468	70	CALIFORNIA, SANTA BARBARA	226
14	RUTGERS	461	71	LAVAL	222
15	INDIANA	442	72	CONNECTICUT	220
16	FLORIDA	435	73	OKLAHOMA	220
17	NEW YORK	433	74	COLORADO	217
18	NORTH CAROLINA	430	75	OREGON	215
19	OHIO STATE	423	76	BROWN	213
20	MINNESOTA	404	77	OKLAHOMA STATE	213
21	PENNSYLVANIA	400	78	TEMPLE	212
22	PRINCETON	398	79	NEBRASKA	211
23	MONTREAL	394	80	SOUTHERN ILLINOIS	211
24	BRIGHAM YOUNG	380	81	ALABAMA	206
25	SOUTHERN CALIFORNIA	376	82	BOSTON COLLEGE	206
26	VIRGINIA	374	83	MIT	206
27	TEXAS A&M	366	84	DELAWARE	203
28	PITTSBURGH	357	85	DARTMOUTH	202
29	UTAH	356	86	HAWAII	202
30	CALIFORNIA, SAN DIEGO	351	87	MISSOURI	200
31	DUKE	342	88	ROCHESTER	200
32	BRITISH COLUMBIA	341	89	WESTERN ONTARIO	199
33	ARIZONA	340	90	HOUSTON	198
34	ARIZONA STATE	340	91	LOUISIANA STATE	195
35	NORTHWESTERN	340	92	TULANE	192
36	JOHNS HOPKINS	337	93	LOUISVILLE	187
37	TEXAS TECH	320	94	IOWA STATE	184
38	ALBERTA	317	95	OHIO	183
39	CHICAGO	308	96	WATERLOO	177
40	GEORGIA	308	97	MASSACHUSETTS	176
41	BOSTON	307	98	QUEEN'S	172
42	EMORY	299	99	SUNY-ALBANY	170
43	IOWA	298	100	HOWARD	167
44	WASHINGTON U.-ST. LOUIS	298	101	WASHINGTON STATE	167
45	KANSAS	292	102	CALIFORNIA, RIVERSIDE	163
46	MICHIGAN STATE	290	103	MCMASTER	163
47	NORTH CAROLINA STATE	288	104	SUNY-STONY BROOK	161
48	TENNESSEE	286	105	VIRGINIA TECH	161
49	NEW MEXICO	283	106	SASKATCHEWAN	159
50	KENTUCKY	281	107	CASE WESTERN RESERVE	150
51	ILLINOIS, CHICAGO	278	108	AUBURN	144
52	MCGILL	276	109	COLORADO STATE	137
53	MIAMI	273	110	RICE	134
54	SOUTH CAROLINA	270	111	GUELPH	131
55	CALIFORNIA, DAVIS	269	112	GEORGIA TECH	125
56	MARYLAND	267	113	KENT STATE	120
57	GEORGETOWN	265			

RANK ORDER TABLE 19
EXPENDITURES FOR ELECTRONIC MATERIALS

RANK	INSTITUTION	VALUE	RANK	INSTITUTION	VALUE
1	PENNSYLVANIA STATE	8,013,755	58	COLORADO STATE	3,034,698
2	COLUMBIA	6,877,655	59	CALIFORNIA, LOS ANGELES	2,988,834
3	TEXAS A&M	6,800,083	60	GUELPH	2,962,912
4	ALBERTA	6,324,354	61	ILLINOIS, CHICAGO	2,918,470
5	WAYNE STATE	5,894,366	62	SOUTH CAROLINA	2,911,445
6	MCGILL	5,844,445	63	MANITOBA	2,878,204
7	WESTERN ONTARIO	5,716,109	64	WASHINGTON STATE	2,844,825
8	NORTHWESTERN	5,550,551	65	MIAMI	2,832,175
9	BRITISH COLUMBIA	5,447,769	66	KENTUCKY	2,725,662
10	NEW YORK	5,407,551	67	VIRGINIA TECH	2,672,867
11	ARIZONA	5,102,800	68	ALABAMA	2,637,709
12	CORNELL	5,016,414	69	SOUTHERN ILLINOIS	2,635,654
13	GEORGIA	4,920,528	70	CALIFORNIA, SAN DIEGO	2,614,758
14	PITTSBURGH	4,905,252	71	RICE	2,610,528
15	JOHNS HOPKINS	4,865,515	72	SASKATCHEWAN	2,597,108
16	YALE	4,864,808	73	SYRACUSE	2,595,344
17	CINCINNATI	4,774,066	74	CASE WESTERN RESERVE	2,590,935
18	OHIO STATE	4,577,236	75	WISCONSIN	2,525,686
19	ARIZONA STATE	4,549,116	76	MIT	2,522,806
20	MINNESOTA	4,509,110	77	OKLAHOMA	2,480,580
21	DELAWARE	4,508,062	78	NEBRASKA	2,343,090
22	TEXAS	4,500,006	79	SUNY-STONY BROOK	2,341,890
23	COLORADO	4,419,033	80	HAWAII	2,264,524
24	FLORIDA	4,398,168	81	OHIO	2,263,405
25	HARVARD	4,370,397	82	MISSOURI	2,259,707
26	WASHINGTON U.-ST. LOUIS	4,315,729	83	PURDUE	2,252,541
27	WASHINGTON	4,282,093	84	GEORGETOWN	2,237,579
28	MARYLAND	4,147,446	85	TENNESSEE	2,213,433
29	VANDERBILT	4,127,501	86	NORTH CAROLINA STATE	2,207,648
30	INDIANA	4,121,438	87	GEORGE WASHINGTON	2,206,549
31	PENNSYLVANIA	4,088,788	88	AUBURN	2,197,192
32	WATERLOO	4,086,189	89	FLORIDA STATE	2,190,436
33	IOWA	4,044,185	90	TEXAS TECH	2,183,952
34	MONTREAL	4,034,046	91	LOUISVILLE	2,171,175
35	NOTRE DAME	4,014,625	92	BOSTON COLLEGE	2,110,452
36	DARTMOUTH	3,920,262	93	BRIGHAM YOUNG	2,098,655
37	CONNECTICUT	3,834,318	94	CALIFORNIA, RIVERSIDE	2,021,541
38	SOUTHERN CALIFORNIA	3,774,856	95	NEW MEXICO	2,004,185
39	CHICAGO	3,755,812	96	MASSACHUSETTS	1,954,626
40	DUKE	3,737,332	97	TULANE	1,945,541
41	TEMPLE	3,727,567	98	KANSAS	1,867,884
42	SUNY-BUFFALO	3,699,962	99	CALIFORNIA, SANTA BARBARA	1,802,251
43	IOWA STATE	3,699,220	100	NORTH CAROLINA	1,716,429
44	MICHIGAN STATE	3,683,654	101	OREGON	1,707,286
45	RUTGERS	3,682,157	102	HOWARD	1,553,660
46	OKLAHOMA STATE	3,594,992	103	KENT STATE	1,553,020
47	EMORY	3,590,667	104	SUNY-ALBANY	1,533,987
48	BROWN	3,565,225	105	HOUSTON	1,526,255
49	CALIFORNIA, IRVINE	3,481,398	106	BOSTON	1,412,656
50	QUEEN'S	3,464,480	107	ROCHESTER	1,357,477
51	PRINCETON	3,371,378	108	LOUISIANA STATE	1,057,160
52	ILLINOIS, URBANA	3,277,386	109	UTAH	726,637
53	VIRGINIA	3,193,088	110	CALIFORNIA, BERKELEY	UA
54	TORONTO	3,153,805	111	GEORGIA TECH	UA
55	CALIFORNIA, DAVIS	3,144,841	112	LAVAL	UA
56	YORK	3,143,591	113	MICHIGAN	UA
57	MCMASTER	3,130,900			

UA – Unavailable

RANK ORDER TABLE 20
ELECTRONIC MATERIALS AS A PERCENT OF TOTAL LIBRARY MATERIALS

RANK	INSTITUTION	VALUE	RANK	INSTITUTION	VALUE
1	WAYNE STATE	76.89	58	CALIFORNIA, RIVERSIDE	37.13
2	WATERLOO	65.96	59	MASSACHUSETTS	37.03
3	WESTERN ONTARIO	64.45	60	RUTGERS	36.72
4	GUELPH	62.76	61	CORNELL	36.42
5	DELAWARE	61.21	62	MINNESOTA	35.90
6	DARTMOUTH	54.16	63	COLUMBIA	35.90
7	OKLAHOMA STATE	53.74	64	NEBRASKA	35.65
8	MCMASTER	52.31	65	MIT	35.33
9	OHIO STATE	51.55	66	NEW MEXICO	34.45
10	MCGILL	51.37	67	WASHINGTON	34.44
11	QUEEN'S	51.19	68	TEXAS	34.22
12	TEMPLE	51.09	69	IOWA	34.08
13	CALIFORNIA, IRVINE	50.27	70	INDIANA	32.52
14	COLORADO	50.21	71	OREGON	32.01
15	SUNY-BUFFALO	50.06	72	SOUTHERN CALIFORNIA	31.84
16	ALBERTA	49.84	73	FLORIDA STATE	31.73
17	WASHINGTON STATE	49.38	74	VIRGINIA	31.67
18	MARYLAND	49.29	75	CALIFORNIA, SANTA BARBARA	31.55
19	CALIFORNIA, DAVIS	49.00	76	CALIFORNIA, SAN DIEGO	31.41
20	COLORADO STATE	48.65	77	PENNSYLVANIA	31.21
21	SYRACUSE	48.50	78	DUKE	30.44
22	NORTHWESTERN	48.11	79	SUNY-ALBANY	29.95
23	CINCINNATI	47.39	80	TULANE	29.81
24	TEXAS A&M	46.99	81	RICE	29.40
25	CONNECTICUT	46.76	82	KENTUCKY	28.96
26	NOTRE DAME	46.34	83	CHICAGO	27.84
27	BRITISH COLUMBIA	46.19	84	BOSTON COLLEGE	27.11
28	ARIZONA STATE	46.06	85	EMORY	26.86
29	SOUTH CAROLINA	45.75	86	MIAMI	26.61
30	PENNSYLVANIA STATE	45.56	87	CALIFORNIA, LOS ANGELES	25.13
31	KENT STATE	45.25	88	LOUISVILLE	25.13
32	MANITOBA	45.23	89	ILLINOIS, URBANA	25.11
33	ARIZONA	45.20	90	NORTH CAROLINA STATE	24.88
34	VIRGINIA TECH	45.09	91	GEORGE WASHINGTON	24.05
35	GEORGIA	44.68	92	GEORGETOWN	23.88
36	BROWN	44.21	93	PRINCETON	23.69
37	WASHINGTON U.-ST. LOUIS	43.29	94	PURDUE	23.61
38	IOWA STATE	43.02	95	TEXAS TECH	23.56
39	MONTREAL	42.38	96	WISCONSIN	23.09
40	ALABAMA	42.04	97	KANSAS	22.41
41	OHIO	41.98	98	OKLAHOMA	22.01
42	AUBURN	41.76	99	TENNESSEE	21.82
43	YORK	41.44	100	ROCHESTER	21.14
44	VANDERBILT	41.41	101	BRIGHAM YOUNG	20.37
45	FLORIDA	41.11	102	HOUSTON	19.87
46	SASKATCHEWAN	40.69	103	BOSTON	17.82
47	HOWARD	39.98	104	TORONTO	15.92
48	JOHNS HOPKINS	39.62	105	HARVARD	15.85
49	MICHIGAN STATE	38.83	106	YALE	15.59
50	MISSOURI	38.74	107	LOUISIANA STATE	15.25
51	HAWAII	38.31	108	NORTH CAROLINA	13.44
52	CASE WESTERN RESERVE	38.19	109	UTAH	9.71
53	ILLINOIS, CHICAGO	37.76	110	CALIFORNIA, BERKELEY	UA
54	SUNY-STONY BROOK	37.67	111	GEORGIA TECH	UA
55	PITTSBURGH	37.53	112	LAVAL	UA
56	NEW YORK	37.41	113	MICHIGAN	UA
57	SOUTHERN ILLINOIS	37.26			

UA – Unavailable

THE ARL MEMBERSHIP CRITERIA INDEX, 2004-05

The ARL Membership Criteria Index is a summary measure of relative size among the university library members of the Association and its sole purpose is to serve as a membership criterion that is quantitative in nature in addition to other qualitative evidence. Both qualitative and quantitative factors support the assessment of research libraries within the Association.

The five variables used in the index were determined by factor analysis of 22 variables originally and represent the elements in which ARL university libraries most resemble one another.[1] They do not attempt to measure a library's services, quality of collections, or success in meeting the needs of users; in comparing any individual library to ARL medians or other ARL members, evaluations must always be made within the context of local goals and characteristics.

The index is calculated using five data elements:

number of volumes held
number of volumes added (gross)
number of current serials received
total operating expenditures
number of professional plus support staff

The index scores are rounded to three decimal places, which may give the appearance of identical scores (and thus ranks) for some institutions; in fact, when the complete principal component score is calculated, it is extended to at least five places. Each institution has a unique result and place in the rank order. Information about index scores and formulae from earlier years as well as excel versions of the index tables are available at: <http://www.arl.org/stats/factor.html>

The formula for 2004-05, based upon the 34 founding members, is:[2]

	. 449766 x log of volumes held
+	. 365633 x log of volumes added (gross)
+	. 476575 x log of current serials
+	. 489405 x log of total library expenditures
+	. 494003 x log of total professional plus support staff
-	27.639982

Decisions on membership issues are being dealt by the ARL Membership Committee which has the responsibility for articulating the principles and tenets of ARL membership, reviews membership criteria, recommends new members to the Board for approval by the full membership, and reviews and makes recommendations on membership status. Principles and procedures for Membership in the Association of Research Libraries are available at http://www.arl.org/stats/qualcov.html . For more information on the ARL Membership Committee agenda, please contact Julia Blixrud, Assistant Executive Director, External Relations (jblix@arl.org). For any questions regarding the methodology used to develop the ARL membership criteria index or any other aspect of the ARL Statistics and Measurement Program, contact Martha Kyrillidou (martha@arl.org).

[1] The 5-variable index was adopted as a membership criterion in May 1987 by the ARL membership. Previous membership criteria were based on a 10-variable index, which had been adopted by the membership in May 1980. The elements used in the 10-variable index were: (1) number of volumes held, (2) number of volumes added (gross), (3) number of microform units held, (4) number of current serials received, (5) expenditures for library materials, (6) expenditures for binding, (7) total salary and wage expenditures, (8) other operating expenditures, (9) number of professional staff, and (10) number of support staff.

[2] "log" refers to the natural logarithm, [ln].

ARL MEMBERSHIP INDEX, 2004-05 - RANKED BY INDEX VALUE

RANK	INSTITUTION	INDEX	RANK	INSTITUTION	INDEX
1	HARVARD	2.51	58	GEORGETOWN	-1.06
2	YALE	1.67	59	LAVAL	-1.12
3	TORONTO	1.31	60	OKLAHOMA STATE	-1.12
4	COLUMBIA	1.27	61	LOUISIANA STATE	-1.12
5	MICHIGAN	1.24	62	COLORADO	-1.13
6	CALIF., BERKELEY	1.16	63	SUNY-BUFFALO	-1.17
7	ILLINOIS, URBANA	0.93	64	NOTRE DAME	-1.19
8	CALIF., LOS ANGELES	0.78	65	BROWN	-1.20
9	CORNELL	0.78	66	NEBRASKA	-1.20
10	TEXAS	0.62	67	DARTMOUTH	-1.28
11	WISCONSIN	0.57	68	BOSTON COLLEGE	-1.29
12	PENNSYLVANIA STATE	0.49	69	WAYNE STATE	-1.30
13	INDIANA	0.42	70	BOSTON	-1.32
14	PRINCETON	0.40	71	MISSOURI	-1.32
15	WASHINGTON	0.32	72	WESTERN ONTARIO	-1.33
16	MINNESOTA	0.21	73	FLORIDA STATE	-1.35
17	NORTH CAROLINA	0.17	74	ROCHESTER	-1.39
18	NEW YORK	0.16	75	SOUTHERN ILLINOIS	-1.39
19	CHICAGO	0.05	76	MIT	-1.43
20	ALBERTA	0.04	77	PURDUE	-1.43
21	PENNSYLVANIA	-0.04	78	YORK	-1.44
22	VIRGINIA	-0.08	79	HAWAII	-1.46
23	PITTSBURGH	-0.08	80	ILLINOIS, CHICAGO	-1.47
24	BRITISH COLUMBIA	-0.13	81	CALIF., SANTA BARBARA	-1.51
25	OHIO STATE	-0.18	82	CALIF., IRVINE	-1.51
26	RUTGERS	-0.20	83	GEORGE WASHINGTON	-1.56
27	FLORIDA	-0.22	84	ALABAMA	-1.57
28	TEXAS A&M	-0.23	85	MASSACHUSETTS	-1.57
29	NORTH CAROLINA STATE	-0.29	86	NEW MEXICO	-1.57
30	JOHNS HOPKINS	-0.30	87	TEMPLE	-1.58
31	IOWA	-0.30	88	IOWA STATE	-1.58
32	DUKE	-0.34	89	LOUISVILLE	-1.59
33	ARIZONA	-0.36	90	SYRACUSE	-1.59
34	NORTHWESTERN	-0.40	91	WASHINGTON STATE	-1.63
35	MONTREAL	-0.42	92	MANITOBA	-1.65
36	GEORGIA	-0.49	93	CALIF., RIVERSIDE	-1.69
37	WASHINGTON U.-ST. LOUIS	-0.49	94	HOUSTON	-1.74
38	SOUTHERN CALIFORNIA	-0.54	95	SUNY-ALBANY	-1.80
39	CALIF., SAN DIEGO	-0.58	96	COLORADO STATE	-1.81
40	EMORY	-0.59	97	SASKATCHEWAN	-1.81
41	ARIZONA STATE	-0.59	98	OHIO	-1.81
42	MCGILL	-0.61	99	VIRGINIA TECH	-1.85
43	KANSAS	-0.62	100	AUBURN	-1.89
44	UTAH	-0.72	101	TULANE	-1.90
45	MICHIGAN STATE	-0.77	102	QUEEN'S	-1.91
46	MARYLAND	-0.85	103	DELAWARE	-1.93
47	CINCINNATI	-0.88	104	GEORGIA TECH	-1.93
48	OKLAHOMA	-0.89	105	OREGON	-1.98
49	TENNESSEE	-0.92	106	SUNY-STONY BROOK	-1.99
50	BRIGHAM YOUNG	-0.94	107	CASE WESTERN RESERVE	-2.01
51	VANDERBILT	-0.96	108	RICE	-2.01
52	MIAMI	-0.97	109	MCMASTER	-2.03
53	CALIF., DAVIS	-0.98	110	WATERLOO	-2.27
54	KENTUCKY	-0.98	111	KENT STATE	-2.47
55	TEXAS TECH	-0.98	112	HOWARD	-2.50
56	SOUTH CAROLINA	-1.04	113	GUELPH	-2.87
57	CONNECTICUT	-1.04			

ARL MEMBERSHIP INDEX, 2004-05 - RANKED BY INSTITUTION

RANK	INSTITUTION	INDEX	RANK	INSTITUTION	INDEX
84	ALABAMA	-1.57	52	MIAMI	-0.97
20	ALBERTA	0.04	5	MICHIGAN	1.24
33	ARIZONA	-0.36	45	MICHIGAN STATE	-0.77
41	ARIZONA STATE	-0.59	16	MINNESOTA	0.21
100	AUBURN	-1.89	71	MISSOURI	-1.32
70	BOSTON	-1.32	76	MIT	-1.43
68	BOSTON COLLEGE	-1.29	35	MONTREAL	-0.42
50	BRIGHAM YOUNG	-0.94	66	NEBRASKA	-1.20
24	BRITISH COLUMBIA	-0.13	86	NEW MEXICO	-1.57
65	BROWN	-1.20	18	NEW YORK	0.16
6	CALIF., BERKELEY	1.16	17	NORTH CAROLINA	0.17
53	CALIF., DAVIS	-0.98	29	NORTH CAROLINA STATE	-0.29
82	CALIF., IRVINE	-1.51	34	NORTHWESTERN	-0.40
8	CALIF., LOS ANGELES	0.78	64	NOTRE DAME	-1.19
93	CALIF., RIVERSIDE	-1.69	98	OHIO	-1.81
39	CALIF., SAN DIEGO	-0.58	25	OHIO STATE	-0.18
81	CALIF., SANTA BARBARA	-1.51	48	OKLAHOMA	-0.89
107	CASE WESTERN RESERVE	-2.01	60	OKLAHOMA STATE	-1.12
19	CHICAGO	0.05	105	OREGON	-1.98
47	CINCINNATI	-0.88	21	PENNSYLVANIA	-0.04
62	COLORADO	-1.13	12	PENNSYLVANIA STATE	0.49
96	COLORADO STATE	-1.81	23	PITTSBURGH	-0.08
4	COLUMBIA	1.27	14	PRINCETON	0.40
57	CONNECTICUT	-1.04	77	PURDUE	-1.43
9	CORNELL	0.78	102	QUEEN'S	-1.91
67	DARTMOUTH	-1.28	108	RICE	-2.01
103	DELAWARE	-1.93	74	ROCHESTER	-1.39
32	DUKE	-0.34	26	RUTGERS	-0.20
40	EMORY	-0.59	97	SASKATCHEWAN	-1.81
27	FLORIDA	-0.22	56	SOUTH CAROLINA	-1.04
73	FLORIDA STATE	-1.35	38	SOUTHERN CALIFORNIA	-0.54
83	GEORGE WASHINGTON	-1.56	75	SOUTHERN ILLINOIS	-1.39
58	GEORGETOWN	-1.06	95	SUNY-ALBANY	-1.80
36	GEORGIA	-0.49	63	SUNY-BUFFALO	-1.17
104	GEORGIA TECH	-1.93	106	SUNY-STONY BROOK	-1.99
113	GUELPH	-2.87	90	SYRACUSE	-1.59
1	HARVARD	2.51	87	TEMPLE	-1.58
79	HAWAII	-1.46	49	TENNESSEE	-0.92
94	HOUSTON	-1.74	10	TEXAS	0.62
112	HOWARD	-2.50	28	TEXAS A&M	-0.23
80	ILLINOIS, CHICAGO	-1.47	55	TEXAS TECH	-0.98
7	ILLINOIS, URBANA	0.93	3	TORONTO	1.31
13	INDIANA	0.42	101	TULANE	-1.90
31	IOWA	-0.30	44	UTAH	-0.72
88	IOWA STATE	-1.58	51	VANDERBILT	-0.96
30	JOHNS HOPKINS	-0.30	22	VIRGINIA	-0.08
43	KANSAS	-0.62	99	VIRGINIA TECH	-1.85
111	KENT STATE	-2.47	15	WASHINGTON	0.32
54	KENTUCKY	-0.98	91	WASHINGTON STATE	-1.63
59	LAVAL	-1.12	37	WASHINGTON U.-ST. LOUIS	-0.49
61	LOUISIANA STATE	-1.12	110	WATERLOO	-2.27
89	LOUISVILLE	-1.59	69	WAYNE STATE	-1.30
92	MANITOBA	-1.65	72	WESTERN ONTARIO	-1.33
46	MARYLAND	-0.85	11	WISCONSIN	0.57
85	MASSACHUSETTS	-1.57	2	YALE	1.67
42	MCGILL	-0.61	78	YORK	-1.44
109	MCMASTER	-2.03			

ARL Statistics Questionnaire and Instructions, 2004-05

ARL STATISTICS QUESTIONNAIRE, 2004-05
INSTRUCTIONS FOR COMPLETING THE QUESTIONNAIRE

General Instructions

Definitions of statistical categories can be found in NISO Z39.7-2004, *Information Services and Use: Metrics & statistics for libraries and information providers--Data Dictionary* (http://www.niso.org/emetrics/current/index.html). ARL has been modifying the interpretation of the standard definitions to address questions posed by by the library staff at various member institutions that complete the survey and with feedback from the ARL Statistics and Assessment Committee (http://www.arl.org/stats/program/meeting.html).

Please do not use decimals. All figures should be rounded to the nearest whole number.

Please respond to every question. If an exact figure cannot be provided, use NA/UA to indicate that the figure is either unavailable or not applicable. If the appropriate answer is zero or none, use **0**.

Although the form allows for data to be entered from both main and branch campuses, an effort should be made to report figures for the main campus only. (The U.S. National Center for Education Statistics, Integrated Postsecondary Education Data System (IPEDS) defines a **branch institution** as "a campus or site of an educational institution that is not temporary, is located in a community beyond a reasonable commuting distance from its parent institution, and offers organized programs of study, not just courses.") If figures for libraries located at branch campuses are reported, please specify which branch libraries are included and which ones are excluded in the notes below.

A **branch library** is defined as an auxiliary library service outlet with quarters separate from the central library of an institution, which has a basic collection of books and other materials, a regular staffing level, and an established schedule. A branch library is administered either by the central library or (as in the case of some law and medical libraries) through the administrative structure of other units within the university. Departmental study/reading rooms are not included.

The questionnaire assumes a fiscal year ending **June 30, 2005**. If your fiscal year is different, please indicate this in the notes below by adjusting the reporting period.

Footnotes. Explanatory footnotes will be included with the published statistics. Provide any notes you may have in the footnotes area at the end of the survey. Reporting libraries are urged to record there any information that would clarify the figures submitted in that line, e.g., the inclusion and exclusion of branch campus libraries. Please make an effort to word your footnotes in a manner consistent with notes appearing in the published report, so that the ARL Office can interpret your footnotes correctly.

Specific Instructions

Question 1. Volumes in Library. Use the ANSI/NISO Z39.7-2004 definition for **volume** as follows:

> *a single physical unit of any printed, typewritten, handwritten, mimeographed, or processed work, distinguished from other units by a separate binding, encasement, portfolio, or other clear distinction, which has been cataloged, classified, and made ready for use, and which is typically the unit used to charge circulation transactions. Either a serial volume is bound, or it comprises the serial issues that would be bound together if the library bound all serials.*

Include duplicates and bound volumes of periodicals. For purposes of this questionnaire, unclassified bound serials arranged in alphabetical order are considered classified. Exclude microforms, maps, nonprint materials, and uncataloged items. If any of these items cannot be excluded, please provide an explanatory footnote.

Include government document volumes that are accessible through the library's catalogs regardless of whether they are separately shelved. "Classified" includes documents arranged by Superintendent of Documents, CODOC, or similar numbers. "Cataloged" includes documents for which records are provided by the library or downloaded from other sources into the library's card or online catalogs. Documents should, to the extent possible, be counted as they would if they were in bound volumes (e.g., 12 issues of an annual serial would be one or two volumes). Title and piece counts should not be considered the same as volume counts. If a volume count has not been kept, it may be estimated through sampling a representative group of title records and determining the corresponding number of volumes, then extrapolating to the rest of the collection. As an alternative, an estimate may be made using the following formulae:

52 documents pieces per foot
10 "traditional" volumes per foot
5.2 documents pieces per volume

Include e-book units, as long as these e-books have been purchased and are owned by your library. If you have access to netLibrary titles as a result of participating in various consortia, **do not** report these e-books as your library's property unless the e-books actually belong to your library. If the books were purchased by a consortium, they may belong to the consortium itself and not to the participating libraries. Report only the number of e-books that belong to your library and are cataloged, classified and made ready for use. Provide a footnote explaining how many e-books you are reporting, preferably by specifying the products and the number of titles in a note.

For information on how to count items housed in remote storage, see the Interim Guidelines for Counting Materials Housed in Library Storage Centers, at http://www.arl.org/stats/arlstat/storage.html .

If either formulas or sampling are used for deriving your count, please indicate in a footnote.

Question 1b. Volumes Added. Include only volumes cataloged, classified, and made ready for use. Include government documents if they have been included in the count of volumes on line 1a. Do not include as part of Volumes Added Gross any government documents or other collections (large gift collections or e-book packages) that were added to the collection as the result of a one time download or addition to the OPAC. Include these items in Volumes Held of the previous year (Line 1a) and provide a footnote explaining the revision of Line 1a.

Include e-book units, as long as these e-books have been purchased and are owned by your library. If you have access to netLibrary titles as a result of participating in various consortia, **do not** report these e-books as your library's property unless the e-books actually belong to your library. If the books were purchased by a consortium, they may belong to the consortium itself and not to the participating libraries. Report only the number of e-books that belong to your library and are cataloged, classified and made ready for use. Provide a footnote explaining how many e-books you are reporting, preferably by specifying the products and the number of titles in a note.

Question 2. Monographic Volumes Purchased. Report number of volumes purchased; do not include volumes received or cataloged. Include all volumes for which an expenditure was made during 2004-05, including volumes paid for in advance but not received during the fiscal year. Include monographs in series and continuations. Include e-books that fit the netLibrary model, i.e., electronic manifestations of physical entities and/or units; provide a footnote explaining how many e-books you are reporting, preferably by specifying the products and the number of titles. If only number of titles purchased can be reported, please report the data and provide an explanatory footnote.

Question 3: Basis of Volume Count. A physical count is a piece count; a bibliographic count is a catalog record count.

Questions 4-5. Serials. Report the total number of subscriptions, not titles. Include duplicate subscriptions and, to the extent possible, all government document serials even if housed in a separate documents collection. Verify the inclusion or exclusion of document serials in Question 5. Exclude unnumbered monographic and publishers' series. Electronic serials acquired as part of an aggregated package (e.g., Project MUSE, BioOne, ScienceDirect) should be counted by title. A **serial** is *a publication in any medium issued in successive parts bearing numerical or chronological designations and intended to be continued indefinitely. This definition includes periodicals, newspapers, and annuals (reports, yearbooks, etc.); the journals, memoirs, proceedings, transactions, etc. of societies; and numbered monographic series.*

In the case of consortial agreements, count under Q4a only those subscriptions to titles for which the library pays directly from its budgeted expenditures reported under Q15b (expenditures for serials). Count under Q4a only those titles and subscriptions for which your library pays. Report other subscriptions that your library receives and does not pay for directly under Q4b (serials received and not purchased). If a purchased subscription includes electronic access to the title, count that subscription twice: once for the print version and once for the electronic version. If serials have been purchased through a consortium whose budget is centrally funded and independent from the library's budget, these serials should be reported under Q4b.

Do not include the full-text serials from such indexing/abstracting products as Wilson Social Sciences Abstracts Full Text, Lexis-Nexis, ABI/INFORM, and other indexes with access to the full text of articles. These full-text titles are counted in in the ARL Supplementary Statistics.

Question 4b. Serials: Not Purchased. Record those serials whose subscriptions were received without purchase for whatever reason. If separate counts of nonpurchased and purchased serials are not available, report only the total number of current serials received on line 4, and report U/A for lines 4a and 4b.

Question 6. Microforms. Report the total number of physical units: reels of microfilm, microcards, and microprint and microfiche sheets. Include all government documents in microform; provide a footnote if documents are excluded.

Question 7. Government documents. Report the total number of physical units (pieces) of government documents in paper format that have not been counted elsewhere. Include local, state, national, and international documents; include documents purchased from a commercial source if shelved with separate documents collections and not counted above. Include serials and monographs. To estimate pieces from a measurement of linear feet, use the formula *1 foot = 52 pieces* and indicate in a footnote that the count is based on this estimate. Exclude microforms and nonprint formats such as maps or CD-ROMs. Adjust line 1a, i.e., last year's Volumes Held, and provide a footnote if you are adding records to the OPAC for government documents previously held but not counted as part of Volumes Held (line 1a).

Question 8. Computer files. Include the number of pieces of computer-readable disks, tapes, CD-ROMs, and similar machine-readable files comprising data or programs that are **locally held as part of the library's collections** available to library clients. Examples are U.S. Census data tapes, sample research software, locally-mounted databases, and reference tools on CD-ROM, tape or disk. Exclude bibliographic records used to manage the collection (i.e., the library's own catalog in machine-readable form), library system software, and microcomputer software used only by the library staff.

Question 9. Manuscripts and archives. Include both manuscripts and archives measured in linear feet.

Question 10. Cartographic materials. Include the numbers of pieces of two- and three-dimensional maps and globes. Include satellite and aerial photographs and images.

Question 11. Graphic materials. Include the number of pieces of prints, pictures, photographs, postcards, slides, transparencies, film strips, and the like.

Question 12. Audio materials. Include the number of pieces of audiocassettes, phonodiscs, audio compact discs, reel-to-reel tapes, and other sound recordings.

Question 13. Film and video materials. Include the number of pieces of motion pictures, videocassettes, video laser discs, and similar visual materials.

Questions 14-20. Expenditures. Report all expenditures of funds that come to the library from the regular institutional budget, and from sources such as research grants, special projects, gifts and endowments, and fees for service. (For question 17, include non-library funds; see instruction Q17.) Do not report encumbrances of funds that have not yet been expended. **Canadian libraries should report expenditures in Canadian dollars.** (For your information, if interested in determining figures in U.S. dollars, divide Canadian dollar amounts by 1.24971, the average monthly noon exchange rate published in the Bank of Canada *Review* for the period July 2004-June 2005.) **Please round figures to the nearest dollar.**

Question 15a. Monographs. Report expenditures for volumes counted on line 2.

Question 15b. Current Serials. Report expenditures for serials counted on line 4a. Exclude unnumbered monographic and publishers' series, and encumbrances.

Question 15c. Other library materials. Include expenditures for all materials not reported in Questions 15a and 15b, e.g., backfiles of serials, charts and maps, audiovisual materials, manuscripts, etc. If expenditures for these materials are included in lines 15a and/or 15b and cannot be disaggregated, please report U/A and provide a footnote. Do not include encumbrances.

Question 15d. Miscellaneous expenditures. Include any other **materials funds expenditures** not included in questions 15a-c, e.g., expenditures for bibliographic utilities, literature searching, security devices, memberships for the purposes of publications, etc. Please list categories, with amounts, in a footnote. **Note:** If your library does not use materials funds for non-materials expenditures—i.e., such expenditures are included in "Other Operating Expenditures"— **report 0, not U/A,** on line 15d.

Question 16. Contract Binding. Include only contract expenditures for binding done outside the library. If all binding is done in-house, state this fact and give in-house expenditures in a footnote; do not include personnel expenditures. (This figure should also be reported in the 2004-05 ARL Preservation Survey, question 7b.)

Questions 17. Salaries and wages. Exclude fringe benefits. If professional and support staff salaries cannot be separated, enter **U/A**, on lines 17a and 17b and enter total staff on line 17.

Question 17c. Salaries and wages: Student Assistants. Report 100% of student wages regardless of budgetary source of funds. Include federal and local funds for work study students.

Question 19. Other operating expenditures. Exclude expenditures for buildings, maintenance, and fringe benefits.

Questions 21-25. Electronic materials expenditures. These items are intended to indicate what portion of your institution's total library expenditures are dedicated to electronic resources and services. Please use the Footnotes to indicate any electronic materials expenditures you believe not to be covered by these questions. **Many expenditures recorded in these questions should have been included in Question 20, total library expenditures.**

Question 21. Computer files. Report expenditures that are not current serials (i.e. are non-subscription, one-time, or monographic in nature) for software and machine-readable materials considered part of the collections. Examples include periodical backfiles, literature collections, one-time costs for JSTOR membership, etc. Expenditures reported here may be derived from any of the following categories: Monographs (Q15a), Other Library Materials (Q15c), Miscellaneous (Q15d), or Other Operating Expenditures (Q19).

Question 22. Electronic Serials. Report subscription expenditures (or those which are expected to be ongoing commitments) for serial publications whose primary format is electronic and for online searches of remote databases such as OCLC FirstSearch, DIALOG, Lexis-Nexis, etc. Examples include paid subscriptions for electronic journals and indexes/abstracts available via the Internet, CD-ROM serials, and annual access fees for resources purchased on a "one-time" basis, such as literature collections, JSTOR membership, etc. Not all items whose expenditures are counted here will be included in Total Current Serials (Question 4) or Current Serial Expenditures (Question 15b).

Q23. Bibliographic Utilities, Networks, and Consortia. Because it is increasingly common for ARL Libraries to enter into consortial arrangements to purchase access to electronic resources, both "Library" and "External" expenditure blanks and instructions are provided. Please use afootnote to describe expenditures that you believe are not covered by the question, or situations that do not seem to fit the instructions.

Q23a. From internal library sources. Report expenditures paid by the Library for services provided by national, regional, and local bibliographic utilities, networks, and consortia, such as OCLC and RLG, unless for user database access and subscriptions, which should be reported in Questions 21 or 22. Include only expenditures that are part of Other Operating Expenditures (Q19).

Q23b. From external sources. If your library receives access to computer files, electronic serials or search services through one or more centrally-funded system or consortial arrangements for which it does not pay fully and/or directly (for example, funding is provided by the state on behalf of all members), enter the amount paid by external bodies on its behalf. If the specific dollar amount is not known, but the total student FTE for the consortium and amount spent for the academic members are known, divide the overall amount spent by your institution's share of the total student FTE.

Q24. Computer hardware and software. Report expenditures from the library budget for computer hardware and software used to support library operations, whether purchased or leased, mainframe or microcomputer, and whether for staff or public use. Include expenditures for: maintenance; equipment used to run information service products when those expenditures can be separated from the price of the product; telecommunications infrastructure costs, such as wiring, hubs, routers, etc. Include only expenditures that are part of Other Operating Expenditures (Q19).

Q25. Document Delivery/Interlibrary Loan. Report expenditures for document delivery and interlibrary loan services (both borrowing and lending). Include fees paid for photocopies, costs of telefacsimile transmission, royalties and access fees paid to provide document delivery or interlibrary loan. Include fees paid to bibliographic utilities if the portion paid for interlibrary loan can be separately counted. Include only expenditures that are part of Miscellaneous Materials Expenditures (Q15d) or Other Operating Expenditures (Q19), and only for those ILL/DD programs with data recorded in Questions 34-35.

Questions 26. Personnel. Report the number of staff in filled positions, or positions that are only temporarily vacant. ARL defines temporarily vacant positions as positions that were vacated during the fiscal year for which ARL data were submitted, for which there is a firm intent to refill, and for which there are expenditures for salaries reported on line 17.

Include cost recovery positions and staff hired for special projects and grants, but provide an explanatory footnote indicating the number of such staff. If such staff cannot be included, provide a footnote. To compute full-time equivalents of part-time employees and student assistants, take the total number of hours per week (or year) worked by part-time employees in each category and divide it by the number of hours considered by the reporting library to be a full-time work week (or year). **Round figures to the nearest whole numbers.**

Question 26a. Professional Staff. Since the criteria for determining professional status vary among libraries, there is no attempt to define the term "professional." Each library should report those staff members it considers professional, including, when appropriate, staff who are not librarians in the strict sense of the term, for example computer experts, systems analysts, or budget officers.

Question 26c. Student Assistants. Report the total FTE (see instruction Q26) of student assistants employed on an hourly basis whose wages are paid from funds under library control or from a budget other than the library's, including federal work-study programs. Exclude maintenance and custodial staff.

Question 27. Number of staffed library service points. Count the number of staffed public service points in the main library and in all branch libraries reported in this inventory, including reference desks, information desks, circulation, current periodicals, reserve rooms, reprographic services (if staffed as a public facility), etc. Report the number of designated locations, not the number of staff.

Question 28. Number of weekly public service hours. Report an unduplicated count of the total public service hours per typical full-service week (i.e., no holidays or other special accommodations) across both main library and branches using the following method (corresponds to IPEDS): If a library is open from 9:00 a.m. to 5:00 p.m. Monday through Friday, it should report 40 hours per week. If several of its branches are also open during these hours, the figure remains 40 hours per week. Should Branch A also be open one evening from 7:00 p.m. to 9:00 p.m., the total hours during which users can find service somewhere within the system becomes 42 hours per week. If Branch B is open the same hours on the same evening, the count is still 42, but if Branch B is open two hours on another evening, or remains open two hours later, the total is then 44 hours per week. **Exclude 24-hour unstaffed reserve or similar reading rooms.** The maximum total is 168 (i.e., a staffed reading room open 7 days per week, 24 hours per day).

Questions 29-30. Instruction. Sampling based on a typical week may be used to extrapolate TO A FULL YEAR for Questions 29 and 30. Please indicate if responses are based on sampling.

Question 29. Presentations to Groups. Report the total number of sessions during the year of presentations made as part of formal bibliographic instruction programs and through other planned class presentations, orientation sessions, and tours. If the library sponsors multi-session or credit courses that meet several times over the course of a semester, each session should be counted. Presentations to groups may be for either bibliographic instruction, cultural, recreational, or educational purposes. Presentations both on and off the premises should be included as long as they are sponsored by the library. Do not include meetings sponsored by other groups using library meeting rooms. Do not include training for library staff; the purpose of this question is to capture information about the services the library provides for its clientele. Please indicate if the figure is based on sampling.

Question 30. Participants in Group Presentations. Report the total number of participants in the presentations reported on line 29. For multi-session classes with a constant enrollment, count each person only once. Personal, one-to-one instruction in the use of sources should be counted as reference transactions on line 31. Please indicate if the figure is based on sampling. Use a footnote to describe any special situations.

Question 31. Reference Transactions. Report the total number of reference transactions. A **reference transaction** is

> *an information contact that involves the knowledge, use, recommendations, interpretation, or instruction in the use of one or more information sources by a member of the library staff. The term includes information and referral service. Information sources include (a) printed and nonprinted material; (b) machine-readable databases (including computer-assisted instruction); (c) the library's own catalogs and other holdings records; (d) other libraries and institutions through communication or referral; and (e) persons both inside and outside the library. When a staff member uses information gained from previous use of information sources to answer a question, the transaction is reported as a reference transaction even if the source is not consulted again.*

If a contact includes both reference and directional services, it should be reported as one reference transaction. Duration should not be an element in determining whether a transaction is a reference transaction. Sampling based on a typical week may be used to extrapolate TO A FULL YEAR for Question 31. Please indicate if the figure is based on sampling.

EXCLUDE SIMPLE DIRECTIONAL QUESTIONS. A directional transaction is an information contact that facilitates the logistical use of the library and that does not involve the knowledge, use, recommendations, interpretation, or instruction in the use of any information sources other than those that describe the library, such as schedules, floor plans, and handbooks.

Questions 32-33. Circulation. For Question 32, count the number of initial circulations during the fiscal year from the general collection for use usually (although not always) outside the library. Do not count renewals. Include circulations to and from remote storage facilities for library users (i.e., do not include transactions reflecting transfers or stages of technical processing). Count the total number of items lent, not the number of borrowers. For

Question 33, report total circulation for the fiscal year including initial transactions reported on line 32 and renewal transactions. Exclude reserve circulations; these are no longer reported.

Questions 34-35. Interlibrary Loans. Report the number of requests for material (both returnables and non-returnables) provided to other libraries on line 34 and the number of filled requests received from other libraries or providers on line 35. On both lines, include originals, photocopies, and materials sent by telefacsimile or other forms of electronic transmission. Include patron-initiated transactions. Exclude requests for materials locally owned and available on the shelves or electronically. Do not include transactions between libraries covered by this questionnaire.

Questions 36. Ph.D. Degrees. Report the number awarded during the 2004-05 fiscal year. Please note that only the number of Ph.D. degrees are to be counted. Statistics on all other advanced degrees (e.g., D.Ed., D.P.A., M.D., J.D.) should not be reported in this survey. If you are unable to provide a figure for Ph.D.s only, please add a footnote.

Question 37. Ph.D. Fields. For the purposes of this report, Ph.D. fields are defined as the specific discipline specialties enumerated in the U.S. Department of Education's Integrated Postsecondary Education Data System (IPEDS) "Completions" Survey. Although the IPEDS form requests figures for all doctoral degrees, only fields in which Ph.D.s are awarded should be reported on the ARL questionnaire. Any exceptions should be footnoted.

Question 38. Instructional Faculty. Instructional faculty are defined by the U.S. Dept. of Education as

> *members of the instruction/research staff who are employed full-time as defined by the institution, including faculty with released time for research and faculty on sabbatical leave.*

Full-time counts generally exclude faculty who are employed to teach fewer than two semesters, three quarters, two trimesters, or two four-month sessions; replacements for faculty on sabbatical leave or leave without pay; faculty for preclinical and clinical medicine; faculty who are donating their services; faculty who are members of military organizations and paid on a different pay scale from civilian employees; academic officers, whose primary duties are administrative; and graduate students who assist in the instruction of courses. Please be sure the number reported, and the basis for counting, are consistent with those for 2003-04 (unless in previous years faculty were counted who should have been excluded according to the above definition). Please footnote any discrepancies.

Questions 39-42. Enrollment. U.S. libraries should use the Fall 2004 enrollment figures reported to the Department of Education on the Integrated Postsecondary Education Data System survey. Please check these figures against the enrollment figures reported to ARL last year to ensure consistency and accuracy. **Note:** In the past, the number of part-time students reported was FTE; the number now reported to IPEDS is a head count of part-time students. Canadian libraries should note that the category "graduate students" as reported here includes all post-baccalaureate students.

FOOTNOTES

Please consult the printed copy of the *ARL Statistics 2003-04* for a copy of last year's footnotes. A pdf version is available at: http://www.arl.org/stats/arlstat/. Explanatory footnotes will be included with the published statistics. Reporting libraries are urged to record in the footnote section any information that would clarify the figures submitted, e.g., the inclusion and exclusion of branch campus libraries (see the "General Instructions" for definition of branch campus libraries). Please make an effort to word your footnotes in a manner consistent with notes appearing in the published report, so that the ARL Office can interpret your footnotes correctly.

Submit the completed questionnaire
by **December 20, 2005.**

For assistance, please e-mail Martha Kyrillidou (martha@arl.org)
or Mark Young (stats-ra@arl.org)
Tel. (202) 296-2296.

Includes:
Main Library _____
Law Library_____
Health Sciences Library_____

ARL STATISTICS 2004-05
WORKSHEET

This worksheet is designed to help you plan your submission for the 2004-05 *ARL Statistics*. The figures on this worksheet should be similar to those in the "Summary" page of your web form, except in cases where data are unavailable. If an exact figure is unavailable, use "NA/UA". If the appropriate answer is zero or none, use "0."

Reporting Institution________________________ Date Returned to ARL ____________

Report Prepared by (name)____

Title ____

Email address ________________________ Phone number____________

Contact person (if different) ____

Title ____

Email address ________________________ Phone number____________

PAGE ONE – VOLUMES

1. Volumes held June 30, 2005

1a. Volumes held June 30, 2004 (1.a) ____________

1b. Volumes added during the year

(i) Volumes added – Gross (1.b.i) ____________

(ii) Volumes withdrawn during year (1.b.ii) ____________

(Net Volumes Added: 1.b.i – 1.b.ii) (1.b) ____________

(Volumes held June 30, 2005: 1.a + 1.b) (1) ____________

2. Number of monographic volumes purchased (2) ____________

3. Basis of volume count is: (3) _____ Physical

_____ Bibliographic

Includes:
Main Library _____
Law Library_____
Health Sciences Library_____

PAGE TWO – OTHER COLLECTIONS

SERIALS

4. Total number of current serials received, including periodicals

4a. Number of current serials purchased (4a) ____________

4b. Number of current serials received but not purchased (4b) ____________
(Exchanges, gifts, deposits, etc. See instructions.)

(Total serials received: 4.a + 4.b) (4) ________________

5. Government documents are included in count of Current Serials? (5) _____ Yes _____ No

OTHER LIBRARY MATERIALS

6. Microform units (6) ________________

7. Government documents not counted elsewhere (7) ________________

8. Computer files (8) ________________

9. Manuscripts and archives (linear ft.) (9) ________________

AUDIOVISUAL MATERIALS

10. Cartographic (10) ________________

11. Graphic (11) ________________

12. Audio (12) ________________

13. Film and Video (13) ________________

Includes:
Main Library _____
Law Library_____
Health Sciences Library_____

PAGE THREE – EXPENDITURES

14. Are the below figures reported in Canadian dollars? (14) ______Yes
______No

15. Total Library Materials Expenditures

15a. Monographs (15a) ________________

15b. Current serials, including periodicals (15b) ________________

15c. Other Library Materials (15c) ________________

15d. Miscellaneous (15d) ________________

(Total library materials: 15.a + 15.b + 15.c + 15.d) (15) ________________

16. Contract binding (16) ________________

17. Total Salaries and Wages

17a. Professional staff (17a) ________________

17b. Support staff (17b) ________________

17c. Student assistants (17c) ________________

(Total salaries and wages: 17.a + 17.b + 17.c) (17) ________________

18. Fringe benefits are included in expenditures for salaries and wages? (18) _____ Yes
_____ No

19. Other operating expenditures (19) ________________

20. Total library expenditures *(15 + 16 + 17 + 19)* (20) ________________

ELECTRONIC MATERIALS EXPENDITURES

21. Computer files *(One-time/monographic purchases.)* (21) ________________

22. Electronic serials (22) ________________

23. Bibliograhpic Utilities, Networks, and Consortia

23a. From internal library sources (23a) ________________

23b. From external sources (23b) ________________

24. Computer hardware and software (24) ________________

25. Document Delivery/Interlibrary Loan (25) ________________

Includes:
Main Library _____
Law Library_____
Health Sciences Library_____

PAGE FOUR – PERSONNEL AND PUBLIC SERVICES

PERSONNEL *(Round figures to nearest whole number.)*

26. Total Staff FTE

26a. Professional staff (26a) ____________

26b. Support staff (26b) ____________

26c. Student assistants (26c) ____________

(Total staff FTE: 26.a + 26.b + 26.c) (26) ________________

STAFFED SERVICE POINTS AND HOURS

27. Number of staffed library service points (27) ________________

28. Number of weekly public service hours (28) ________________

INSTRUCTION

29. Number of library presentations to groups (29) ________________

29a. Figure based on sampling? (29a) ______Yes ______No

30. Number of total participants in group presentations reported in line 29
(30) ________________

30a. Figure based on sampling? (30a) ______Yes ______No

REFERENCE

31. Number of reference transactions (31) ________________

31a. Figure based on sampling? (31a) ______Yes ______No

Includes:
Main Library _____
Law Library_____
Health Sciences Library_____

PAGE FIVE – PUBLIC SERVICES AND LOCAL CHARACTERISTICS

CIRCULATION

32. Number of initial circulations (excluding reserves) (32) ________________

33. Total circulations (initial and renewals, excluding reserves) (33) ________________

INTERLIBRARY LOANS

34. Total number of filled requests provided to other libraries (34) ________________

35. Total number of filled requests received from other libraries or providers (35) ________________

Ph.D. DEGREES AND FACULTY

36. Number of Ph.D.s awarded in FY2004-05 (36) ________________

37. Number of fields in which Ph.D.s can be awarded (37) ________________

38. Number of full-time instructional faculty in FY2004-05 (38) ________________

ENROLLMENT – FALL 2004
(Line numbers refer to IPEDS survey form.)

39. Full-time students, undergraduate and graduate (39) ________________
(Add line 8, columns 15 & 16, and line 14, columns 15 & 16.)

40. Part-time students, undergraduate and graduate (40) ________________
(Add line 22, columns 15 & 16, and line 28, columns 15 & 16.)

41. Full-time graduate students *(Line 14, columns 15 & 16.)* (41) ________________

42. Part-time graduate students *(Line 28, columns 15 & 16.)* (42) ________________

FOOTNOTES

On the web form, you will be able to add footnotes to individual questions, as well as footnotes that apply to your entire institution. Please provide any information which would clarify the figures submitted, e.g., the inclusion of branch campus libraries or any special projects which might cause radical increases or decreases from the 2003-04 data. **The form will not be able to provide your institution's 2003-04 footnotes**; please use the footnotes published in the *ARL Statistics 2003-04* for comparison if necessary. Please make an effort to word your footnotes in a manner consistent with notes appearing in the published report, so that the ARL Office can interpret your footnotes correctly.

Submit the completed questionnaire by **December 20, 2005.**

For assistance, please e-mail Martha Kyrillidou (martha@arl.org)
or Mark Young (stats-ra@arl.org)
Tel. (202) 296-2296.

FOOTNOTES TO THE ARL STATISTICS, 2004-05

Footnotes may also include errata and corrections to data from prior years not previously reported. Numbers in parentheses refer to columns in Library Data Tables and to Questionnaire numbers.

QUESTION NUMBER	NOTES
ALABAMA	
	All figures are as of September 30, 2005, and include University Libraries (Gorgas, Business, Education, Special Collections, Science & Engineering), Map, Social Work, and Communications libraries. Volumes held September 30, 2004 revised to 2,466,978 to include pre-1996 government documents volumes added to catalog during 2004-05.
1	Includes government documents since 1996 and a portion of the pre-1996 government documents collection.
2	Law Library purchased a PDF set with one-time funding.
6	Includes government documents.
15.b	Decrease reflects change in renewal cycle.
15.c	Includes only Map Library; other libraries do not separate expenditures for audiovisual materials and microforms.
15.d	Includes Center for Research Libraries membership.
17.b	Increase due to additional staff, hiring at higher salary, and raises.
19	Includes bibliographic utilities, memberships (except CRL), and literature searching.
21	Includes purchase of e-books; in 2003-04 also included purchase of backfiles.
22	Increase due to moving major publisher' packages from print and electronic to electronic only.
26.a	Reclassification moved staff from professional to support.
ALBERTA	
	All figures are as of March 31, 2005. Volumes held June 30, 2004 revised to 6,027,270 to include e-book titles held as at end of March, 2004.
1.b.i	Includes 81,409 e-book titles added in 2004-05. Excludes 1,811 e-book titles purchased for COPPUL.
2	Past figures were inaccurately reported. Precise number of monographs purchased from 2000-01 onward is unavailable; all monographs purchased are included in gross volumes added (1.b.i)
4	Includes government documents.
7	Included in Microform Units (6).
15-25	Expenditures as reported in Canadian Dollars: (15a) $5,368,076; (15b) $9,794,761; (15c) $0; (15d) $694,766; (15) $15,857,603; (16) $117,796; (17a) $5,743,088; (17b) $9,166,184; (17c) $1,001,858; (17) $15,911,130; (19) $1,567,792; (20) $33,454,321; (21) $969,687; (22) $6,933,922; (23a) U/A; (23b) U/A; (24) U/A; (25) U/A.
15.c	Cannot be disaggregated from monograph expenditures (15a) and serial expenditures (15b).
15.d	Includes GST (Goods and Services Tax), postage, courier, and supplies expenditures, as well as membership expenditures not already paid through the materials budget.
17	Excludes Fringe benefits.
21	Includes one-time monograph fund purchases of electronic periodical backfiles.
26.c	Figure based on actual hours worked. Prior to 2000-01, figures were based on hours hired to work.
33	Includes Journal Loans.

QUESTION NUMBER	NOTES
35	Figure was unavailable for 2003-04.
36, 37	Data can vary from year to year.
ARIZONA	
	Volumes held June 30, 2004 revised to 5,271,857 to include 70,792 volumes added to base for purchased e-book collections.
1.b.i	Includes 67,404 regular additions plus 26,426 e-books from sources like IEEE and SPIE, and NetLibrary owned titles.
4	Main library reported (4a) 11,396; (4b) 11,739; others U/A.
23.a	Includes Health Sciences Library only.
ARIZONA STATE	
	Includes University Libraries, ASU West, ASU Polytechnic, and Law Library. Volumes held June 30, 2004 revised to 4,091,558 to include: an increase of 32,883 due to corrections for over reporting in 2003-04; addition of three large e-book record loads into online catalog during 2004-05; correction for double reporting of a netLibrary load; an internal spreadsheet calculation error in 2003-04; 20,125 e-brary records loaded May/June 2005; 15,350 additional Eighteenth Century Collections Online (ECCO) records loaded March 2005; 627 Knovel records loaded May 2005.
15.d	Includes: Conservation/Preservation ($19,101); Memberships for purposes of publications ($104,922); ILL/Document delivery ($17,505); uncataloged office copy language dictionary ($63).
AUBURN	
	All figures are as of September 30, 2005. Volumes held June 30, 2004 revised to 2,903,974 due to purchase of ECCO.
1.b.ii	Decrease due to withdrawal of duplicate journals at VetMed Library.
2	Figure reflects purchasing patterns for collection managers during 2004-05.
4.a	Figure reflects decrease in on-line only serials.
4.b	Decrease due to exchange program that doesn't require holding of additional periodicals.
7	Based on shelf measurement rather than bibliographic records.
12	Added NAXOS music collection.
15.a	Decrease in approval shipments.
15.c	Indexes and abstract databases changed to this category.
23.b	Estimated number based on cost if not provided by consortia.
24	Expenditures for new Digital Resources Lab outfitting included in 2003-04.
25	Data in (23a) was erroneously included here in previous years.
31	Added additional service point and improved data collection.
35	Decrease in requests due to improved library holdings.
42	More graduate student shifting from full-time to part-time.
BOSTON	
4, 7, 11, 23.b	Main and law library data unavailable.
6, 8	Excludes special collections (reported with main library statistics).
15.d	Includes main library only.
BOSTON COLLEGE	

QUESTION NUMBER	NOTES
	Includes Bapst Art Library, John J Burns Library of Rare Books and Special Collections, Educational Resource Center, and the School of Social Work Library.
2, 8, 21	O'Neill Library does not collect this information.
4, 4.a	2003-04 figure revised to (4a) 35,896; (4) 49,074 to reflect proper counting of e-journals.
8	Figure reflects the Law Library only.

BOSTON PUBLIC

	Volumes held June 30, 2004 revised to 7,443,759.

BRIGHAM YOUNG

	All figures reflect the status of collections and expenditures as of December 31, 2004.
8	Figure includes Law Library only.
9	Reassessment of Special Collections has resulted in a dramatic increase in linear feet being occupied by manuscripts and archives.

BRITISH COLUMBIA

	All figures are as of March 31, 2005 and include the following libraries: Applied Sciences, Archaeology, Architecture, Asian, Biomedical Branch, B'nai B'rith, Botanical Garden, Integrated Computer Systems Research, Centre for Teaching and Academic Growth, Coolie Verner (Adult Education), Crane Resources Centre, David Lam Management Library, Earth & Ocean Sciences, Education, Fine Arts, First Nations House of Learning, Geographic Information Centre, Hamber, Language Education Resource Centre, Law, MacMillan, Main, Mathematics, Music, Museum of Anthropology, Psychiatry RR, Psychoed IC, Rare Books & Special Collections, Regent Carey College, Robson Square, Science & Engineering, St.Mark's College, St. Paul's Hospital, TRIUMF, University Archives, Vancouver School of Theology, Walter C. Koerner Library, Woodward. Medical and Law Library statistics cannot be disaggregated from the Main statistics. Volumes held June 30, 2004 revised to 5,437,678 to include the purchase of 229,837 e-books in 2004-05.
1.b.i	Decrease reflects a change in major monograph vendor and processing methods.
15-25	Expenditures as reported in Canadian dollars: (15a) $2,461,684; (15b) $10,537,532; (15c) $767,193; (15d) $972,890; (15) $14,739,299; (16) $191,702; (17a) $7,438,156; (17b) $7,307,198; (17c) $1,294,241; (17) $16,039,595; (19) $3,044,360; (20) $34,014,956; (21) $262,768; (22) $6,545,364; (23a) $44,367; (23b) $0; (24) UA/NA; (25) $212,197.
22	Transition to online program.
33	Decrease reflects the different way in which Voyager manages renewals, and a change in loan period for grad students decreased number of renewals.
38	Excludes deans and faculty on disability benefits.

BROWN

	Includes Rockefeller, Sciences, John Hay, Orwig Music, Artslide, and John Carter Brown libraries. Medical Library statistics cannot be disaggregated from the main because the Medical collection is an integral part of the Sciences Library. Volumes held June 30, 2004 revised to 3,383,362 to include 156,094 electronic books.
3	Basis of volume count is both physical and bibliographic since electronic books have been added.
19	2003-04 figure revised to $2,007,720.

CALIFORNIA, BERKELEY

	Volumes held June 30, 2004 revised to 9,828,970.
4	Increase due to an increase in the number of print subscriptions, and also due to shared subscriptions being included for the first time.

QUESTION NUMBER	NOTES
15.c	Data are included in monograph expenditures (15a).

CALIFORNIA, DAVIS

	Volumes held June 30, 2004 adjusted to reflect the true number of volumes on hand. A conversion of government documents to volumes using the ARL guidelines was initially reported in 1991-92; however, in subsequent years a combination of reporting both pieces and volumes has occurred. A thorough review of all records pertaining to government documents has been completed and in March of 2005 the correct number of volumes and government document pieces as of June 30, 2004 was determined. As of 2004-05 all government documents residing in the Government Information unit will be reported as volumes.
4	Increase due to the addition of new electronic serials titles.
7	Figure adjusted to reflect the true number of items on hand as of June 30, 2004. As of 2004-2005 all government documents residing in the Government Information unit will be reported as volumes.
15.b	The Library prepaid serials for the Main Library in the amount of $503,495 in 2003-04.
17.c	Expenditures in this category dropped by 27.78% due to a reduction in project oriented activities (i.e. collections shifts) and a reduction in the need for typical student employment tasks in public services such as the re-shelving of books due to a reduction in the number of circulations of physical items.
33	As the number of items available to patrons online increases, the circulation of printed material decreases.

CALIFORNIA, IRVINE

1.b	Fewer backlog and gift items were cataloged; government publications also decreased.
17.a	Accumulation of unfilled vacancies due to tight budgets in the previous two years.
17.c	Streamlining and reorganization led to less reliance on student employees.
21, 23.b, 24, 38	2003-04 figures were incorrect.

CALIFORNIA, LOS ANGELES

	Volumes held June 30, 2004 revised to 6,812,576 to reflect a unit added for 2003-04.
1.b.i	Decrease due to migration preparation and the transition to a new ILS.
1.b.ii	There was an increase in the number of items discarded in lieu of deposit.
6	2003-04 figure revised to 6,135,957.
21-25	Law and Health Sciences figures cannot be disaggregated from the Main Library.
24	Decrease due to the cyclical nature of these purchases.
34, 35	Includes Main Library only.

CALIFORNIA, RIVERSIDE

4.b	Increase due to centrally funded electronic serials for the UC Digital Library.
22	2003-04 figure revised to $2,067,533 due to centrally allocated funds that were inadvertently omitted.
32, 33	2003-04 figures revised to (32) 194,291; (33) 265,554; due to reserve circulations that were inadvertently included.

CALIFORNIA, SAN DIEGO

	UCSD has several libraries on the main campus which, for this survey, are grouped under the name of Geisel Libraries. This includes: The Social Sciences and Humanities Library, The Arts Libraries, The Science and Engineering Library, The Scripps Institution of Oceanography Library, The International Relations and Pacific Studies Library, The Mandeville Special Collections Library

QUESTION NUMBER	NOTES
	and the Center for Library and Instructional Computing Services. The Biomedical Library and Medical Center Library statistics are combined for the Health Sciences Library Survey. Volumes held June 30, 2004 revised to 3,071,461 to include 68,826 e-books.
1.b.i	Includes 34,612 e-books added in 2004-05.
2	Excludes 10,470 e-books purchased for all libraries, including the Health Sciences.
11	Withdrew 31,000 slides from the libraries' collections.
15.b	Includes monographs and publisher series. Includes all formats. Includes electronic databases subscribed to annually.
21	Increase due to the addition of a back run of Web of Science.
27	Service points closed/merged to accommodate temporary move into the main library building during construction of the Biomedical Library: maps to reference; biomedical to periodicals.
32, 34	2003-04 figures revised to (32) 495,031; (34) 60,850.

CALIFORNIA, SANTA BARBARA

4.b	Figure reflects proper counting of e-journals received in licensed packages through a consortium.
15.c	Included in monograph expenditures (15a).
23.a	Decrease due to prepayment of OCLC charges for 2005 that were made in 2004.

CASE WESTERN RESERVE

1.b.ii	Major weeding project undertaken at the School of Applied Social Sciences Library in conjunction with reclassification of the collection from Dewey to L.C.
8	Major weeding of out-of-date and unusable formatted materials at the School of Applied Social Sciences library.
15.d	Includes: Dialog ($632); CNI membership ($5,800); ICPSR membership ($12,730); ARL membership ($20,150); SPARC membership ($5,000); TEI membership ($1,000); CLIR membership ($3,500); Theses ($17,592); Council on Social Work Education membership ($380).

CENTER FOR RESEARCH LIBRARIES

15.a, 15.b	In 2004-05 CRL stopped early renewal of serial subscriptions for the dissertation exchanges. The cost of these subscriptions had always been counted as the cost to receive the dissertation monographs. However, since early renewal was done in 2004 for 2005, monographs expenditures for 2005 appear to have decreased.

CHICAGO

	John Crerar Library is a central science library with a biomedical collection for health sciences; many numbers are percentages as a result. Data for the ARL survey are matched as closely as possible to data compiled and reported for the Association of Academic Health Sciences Libraries' (AAHSL) Annual Statistics of Medical School Libraries in the United States and Canada. Volumes held June 30, 2004 was adjusted to 7,205,924 to include 81,545 volumes that were undercounted during the period 2000-04.
2	Approximately 66% of this total represents a title count.
8-10, 12, 13	Excludes unprocessed and partially processed items.
21, 22	The increase in computer files and decrease in subscriptions represents an emphasis on backfiles and other one-time purchases, a shift toward the purchase of packages rather than subscriptions when that option exists and continued refinement of our internal categorizations.
23.a	Includes standard ongoing operating costs only.
23.b	Figure is estimated.

QUESTION NUMBER	NOTES
24	2003-04 figure reflected compensation savings used to support computer expenditures.
35	Includes 411 requests filled by the Center for Research Libraries.
40, 42	Prior to 2004-05, part-time students were reported to IPEDS as FTEs; now, they are reported as a head count.
CINCINNATI	
	Includes the University Libraries, as well as Law and Medical. Excludes Raymond Walters College and Clermont College. Volumes held June 30, 2004 revised to 3,051,048 to reflect adjustments in retro cataloging.
5	The majority of government documents are not counted in the count of Current Serials, though some have been classed in the LC collection and are considered "regular" periodicals.
17.a	Increase reflects a change in the definition of "professional staff" in one unit.
21	Increase due to the purchase of major backfile collections.
22	Decrease due to some payments to the consortium, which were paid in 2003-04.
23.a	Increase due to various fiscal year patterns.
26.c	Budget loss explains the drop in FTE student assistants.
34	Figure includes activity for the University's two two-year colleges: Raymond Walters College and Clermont College. Decrease due to a change in lending priority within the OhioLINK consortium; institutions change into and out of position as priority lender in different years.
35	Figure includes activity for the University's two two-year colleges: Raymond Walters College and Clermont College.
CANADA INSTITUTE FOR SCIENTIFIC AND TECHNICAL INFORMATION	
	All figures are as of March 31, 2005 and include: 11 regional CISTI sites, in Victoria, Vancouver, Winnipeg, Saskatoon, Montreal, Boucherville, London, Halifax, New Brunswick, Charlottetown and St. John's; and 6 Ottawa-based NRC Information Centres staffed by CISTI Information Specialists in addition to the main CISTI building.
15-25	Expenditures as reported in Canadian dollars: (15a) $536,500; (15b) $10,702,057; (15c) $879,790; (15d) $68,209; (15) $12,186,556; (16) UA/NA; (17a) UA/NA; (17b) UA/NA; (17c) UA/NA; (17) $19,367,085; (19) $13,315,444; (20) $44,859,000; (21) UA/NA; (22) $1,911,793; (23a) UA/NA; (23b) UA/NA; (24) $3,290,178; (25) UA/NA.
29	Data not collected in 2005. Figures for question (29) and (30) have matched in past years, but reflect the number of participants.
COLORADO	
3	Volume count is 90% bibliographic and 10% physical.
15.d	Includes: Consortia ($147,003, consisting of Colorado Alliance of Research Libraries membership [$88,566], Center for Research Libraries membership [$50,337], Greater Western Library Alliance membership [$8,100]); Membership ($25,205, consisting of Inter-university Consortium for Political & Social Research [$12,730], ARL [$6,500], CNI (Coalition for Networked Information [$5,975]); Korean cataloging records ($1,718); Marcive elec. shipping lists and labels ($3,633); III LC authority database ($5,900); III maintenance ($94,200); Interlibrary loan ($8,798); Storage of archival materials ($27,360); Pinhole camera ($153); Travel to view gift collection ($359).
19	2003-04 figure should be reduced by $100,000 cash transfer to copy center auxiliary which was double- counted.
29	A couple of departments have revised their methods of counting presentations to better reflect the ARL instructions. This lowered the figure significantly in comparison to previous years.
31	Reflects more accurate counts now uniformly based on ARL definition of reference transactions.

QUESTION NUMBER	NOTES
41	Law students (first-professional) were omitted from full and part-time grad students in Fall 2003, thus 2003-04 figures should be revised to (41) 2,121; (42) 3,927.
COLORADO STATE	
4.a	Change is due to interpretation of what is now included in this item.
15, 15.a, 20	Changes due to completion of flood related projects.
21	Figure reflects shift to online resources.
COLUMBIA	
	Includes two Columbia University affiliates (not branch libraries): Barnard College Library, which was also included in 2003-04, and Teachers College which is included in 2004-05 for the first time. Volumes held June 30, 2004 revised to 9,085,859 to include 5,169 Oral History volumes.
2	Number of monographic volumes purchased includes 159,844 e-books, some of which were purchased in prior years but cataloged in 2004-05.
8	Excludes Butler Library.
14	All Columbia expenditures and Personnel data (FTE) exclude Electronic Publishing Initiative at Columbia (EPIC).
15.d	Includes costs associated with maintenance of collection (e.g., shipping, supplies, relocation, memberships)
16	Decrease reflects re-categorization of expenses.
CONNECTICUT	
	Includes Art & Design; Avery Point; Dodd Research Center; Greater Hartford; Map; Media; Music & Dramatic Arts; Pharmacy; Stamford; Torrington; and Waterbury libraries.
4	Increase due to a change in the state-provided electronic serials package.
15.b, 15.c	2003-04 figures were incorrect.
21	Increase due to an internal coding change.
CORNELL	
	Includes the Adelson, Africana, Annex, Engineering, Fine Arts, Hotel, Industrial & Labor Relations, Kroch (Rare & Mss. Collections), Law, Management, Mann/Entomology, Mathematics, Music, Olin/Kroch (Asia Collections), Physical Sciences, Uris, and Veterinary Libraries on the Ithaca campus, and the New York Agricultural Experiment Station Library in Geneva, NY. Excludes the Medical College Library in New York, NY. Volumes held June 30, 2004 revised to 7,537,145 to include corrections to Annex volume counts, and 240,986 additional e-books, including: Early English Books Online (68,463); Eighteenth Century Collections Online (136,209); Evans Early American Imprints Digital Collection (36,303); NetLibrary (11 ® excludes 12,021 NetLibrary titles Cornell has perpetual access to with a consortium).
1	Decrease largely reflects the completion in 2003-04 of the backlog reduction project. To a much smaller degree, it may also reflect a decrease in the number of unit orders placed due to the performance of the dollar.
1.b.i	Decrease reflects the completion in 2003-04 of a project to eliminate the cataloging backlog; the process of transitioning from acquiring print and electronic to electronic-only serials where appropriate; a decrease in the number of unit orders placed due, in some part, to the performance of the dollar; and a reduction in the acquisition of multiple copies.
4	Includes some unnumbered monographic series. Estimate of e-serials does not exclude non-current titles packaged with current titles.
9	Figure is in cubic feet.
15.a	Decrease due to the fact that 2003-04 figure included funds for the large purchase of the

QUESTION NUMBER	NOTES
	Huntington Free Library Native American Collection.
15.d	Includes expenditures for shipping.
22	Figure is estimated; increase due to the separation of expenditures for those most expensive e-serials, which were previously characterized as print expenditures.
26	Excludes staff in positions that were temporarily vacant. Includes professional, support and student staff on grants and projects.
31	One library could not report, and another was relocated during a renovation for half a year. Two libraries' figures were based on sampling.
33	Decrease due to the fact that reserve transactions were excluded for the first time. Includes Annex material charges made at campus libraries, which were mistakenly excluded from the 2001-02 through 2003-04 figures. As in the past, includes ILL and some staff transactions.
40, 42	Employee degree only.
DARTMOUTH	
1.b	Increase due to an abnormally small amount of weeding of printed volumes in 2004-05.
2	Estimate based on number of payment lines; includes serial backfiles.
4.a	May be high for electronic journals, where free titles are estimated.
4.b	Electronic portion of this is a (low) estimate, based on USGPO serials and free packages identified in Serials Solutions.
17	2003-04 figures revised to (17a) $3,518,762; (17b) $3,372,468; (17c) $480,808; (17) $7,372,038.
19	2003-04 figure revised to $1,710,768.
20	2003-04 figure revised to $16,387,776.
DELAWARE	
9	Excludes the University Archives which is not administered by the Library.
20	Excludes $553,000 grant for the provision of electronic library resources to public schools.
21	Decrease reflects special funding support in 2003-04 for electronic resources from Wiley Interscience.
23.a	Decrease reflects a pre-payment made to PALINET for OCLC in 2003-04.
24	Increase reflects the purchase of equipment for the library' institutional repository.
26	Excludes professional and support staff in Information Technologies assigned to the Library.
DUKE	
	Volumes held June 30, 2004 revised to 4,781,425 due to the previous omission of e-book titles.
1.b.i, 2	Decrease due to implementing new workflows in a new ILS.
23.a, 23.b	Decrease reflects a correction to inaccurate responses submitted in 2003-04.
EMORY	
	All figures are as of August 31, 2005, and include Oxford College Library and the Theology Library. Volumes held June 30, 2004 revised to 3,028,303 to include 136,209 e-books (Eighteenth Century Collections Online) added to the online catalog during 2005.
4	Theology Library serials of 1,545 are not broken down into purchased (4a) and received, but not purchased (4b). Main Library and Oxford were broken down as follows: Main: (4a) 13,636; (4b) 13,652. 2003-04 figures were overly inclusive, containing titles in aggregators and received through GALILEO.

QUESTION NUMBER	NOTES
7	Totals for past few years had been reported as if volumes rather than pieces. 2004-05 figure has been converted to piece count as per instructions.
15.d	Includes Law and Health Sciences libraries only; Main Library data unavailable.
21	Previous years' totals have been significantly under-reported by not including electronic backfiles.
22	Figure reflects the change in billing for primary format of many serial subscriptions, now having electronic as primary format with print as the add-on.
24	Increase due to increase in computer hardware and software purchased for grant-funded projects.
32	Includes Music/Media Library and Chemistry Library as well as Theology (no data was reported for Oxford).
FLORIDA	
4.a	Increase due to first year electronic serial counts being available and reported.
33	Main library closed for renovation.
38	Figure reflects UF Common Data Set, which may not be comparable to previous years.
FLORIDA STATE	
	Includes Strozier Library (Main), Dirac Science Library, College of Law Library, College of Medicine Library, Allen Music Library, Goldstein Information Science Library, Panama City ARC, Panama Branch Campus Library, and Ringling Museum Library.
1.b	Decrease due to the purchase of e-books in 2003-04 that was not repeated in 2004-05.
GEORGE WASHINGTON	
1.b	Decrease due to a major weeding project in 2003-04.
1.b.ii	Main Library withdrew 8,204 in 2004-05 during inventory.
2	Eckles library switched approval vendors; the transition delayed purchase and receipt of monographs.
4	Main Library erroneously reported this figure as 12,005 for a 2003-04 correction. In 2004-05, Main Library increased access to the number of electronic journals via SFX, taking advantage of "free or nearly free with print."
6	Main Library did not count individual rolls of microfilm and individual sheets of microfiche for government document serials prior to 2004-05. Other serials on microfilm and microfiche were counted by individual titles. This is being resolved for 2005-06.
10	The large number in 2003-04 was due to the Main Library clearing out a large backlog of unprocessed maps.
15.b	Main Library used a new methodology to run a retrospective report on serials expenditures (paper and electronic), which resulted in the 2003-04 total being revised to $2,951,733. 2004-05 figure includes the cost of FirstSearch and Lexis-Nexis.
15.c	For 2004-05, Media expenditures for the Main Library were aggregated in with the monograph expenditures (15a).
15.d	Main Library's expenditures for FirstSearch and Lexis-Nexis are counted in serials and electronic expenditures (15b).
19	Includes an expenditure of $358,157 for cost recovery, special projects and grants, and temp services staff. FTE for temp services staff is reported in (26).
21	Includes mostly one-time purchases and excludes serials.
24	In 2004, Main Library implemented a spending freeze on IT purchases to address a budget shortfall. The freeze on IT purchases ended in 2004-05.

QUESTION NUMBER	NOTES
26.a	Main Library figure includes one grant-funded FTE librarian.
26.b	Main Library includes 9 FTE contract and grant cost recovery staff.
29-31	Data unavailable for Law Library.
GEORGIA	
	Includes Main, Science, and Law libraries. Volumes held June 30, 2004 revised to 4,114,611 due to the addition of several e-book packages.
24	2003-04 figure did not include Main Library.
GEORGIA TECH	
1.a	Includes government documents that are catalogued, classified, and housed separately.
24	Figure reflects replacing all computers in the library's main lab.
26.a, 26.b	Decrease due to an internal re-interpretation of the classifications.
33	Data previously available to compute this figure are no longer available following transfer of circulation, bibliographic, and other data to integrated library system Voyager.
GUELPH	
	All figures are as of April 30, 2005.
1	Collections of electronic books that the Library "owns" but has yet to catalogue and make ready for use are not included in this figure. One example would be ◎Eighteenth Century Collections Online."
1.b	Figure reflects a move to electronic periodicals, and space constraints (especially in the Library's storage facility).
4.a	Decrease from 2003-04 figure is due to a revision in the method of determining the number of print serial subscriptions; the Library is adding to its serial holdings, especially in the area of electronic journals, and no cancellation project has been undertaken.
4.b	This is the number of free electronic journals available through the Library's web site.
15-25	Expenditures as reported in Canadian dollars: (15a) $1,295,621; (15b) $3,809,019; (15c) $473,521; (15d) $322,209; (15) $5,900,370; (16) $31,375; (17a) $2,613,844; (17b) $3,102,668; (17c) $272,734; (17) $5,989,246; (19) $1,912,197; (20) $13,833,188; (21) $430,868; (22) $3,271,913; (23a) $34,304; (23b) $0; (24) $544,143; (25) $151,447.
15.c	Includes one-time payments from the serials ledger (e.g. for periodical backfiles) and those related to electronic books (either not owned by the Library or not yet catalogued).
19	Increase due to payments made under the University's "Voluntary early retirement and resignation" program.
26	Decrease due to the fact that some staff took advantage of the University's "Voluntary early retirement and resignation" program.
29	Increase due to the strategic importance given to promoting information literacy.
31	Includes all types of reference interactions: in-person, phone, FAX, email and virtual reference.
32, 33	Excludes circulation counts for materials in Archives and Special Collections.
36	This is the number of PhDs awarded between November 1, 2004 and October 31, 2005.
38	Figure reflects relevant faculty as of September 2004.
42	Increase is not thought to be significant, as the number of part-time graduate students is quite small.
HARVARD	

QUESTION NUMBER	NOTES
	Volumes held June, 30 2004 revised to 15,377,413.
1	Decreased volume count due to the closing of two collections, also reflected in the number of volumes withdrawn. Includes some government documents.
4	Includes some government documents; excludes some international government documents.
15.a	Anomalies in monograph expense are the result of changes in accounting practice and reduced expenditures due to budgetary reductions.
15.c	An increase in expenditures for other library materials reflects rising costs overall and the shift from paper to digital resources.
15.d, 19	Increases due to a change in accounting practice.
21	Increased expenditures for computer files reflect transition from paper to electronic journals.
23.a	Increase reflects several one-time expenditures.
24	Increase attributed to changes in accounting for internal services, some renovation activity, and upgrades.
25	Decrease reflects fewer transactions in addition to changes in accounting system.
HAWAII	
	Excludes the Law and Health Sciences libraries.
1.b.ii	U-Hawaii is still recovering from the flood of October 2004 and is not yet able to factor total loss into the count of withdrawn and lost-in-flood volumes.
6, 10	Figures are the same as in 2003-04. The count of items lost in the flood of October 2004 has not yet been determined.
9	Figure reflects a more accurate count following creation of separate Archives and Manuscripts Department within the library.
26.a	2003-04 figure revised to 70.
HOUSTON	
	All figures are as of August 30, 2005 and include William R. Jenkins Architecture & Art Library, Music Library, Weston A. Pettey Optometry Library, Pharmacy Library, and the O'Quinn Law Library. Volumes held June 30, 2004 revised to 2,317,288 to include the e-book packages LION (13,578 volumes) and NetLibrary (46,847 volumes). In addition, the UH Law Library had a significant decrease in the number of volumes added.
15.b, 15.c	In 2003-04, approximately $740,000 was incorrectly reported under Other Library Materials (15c). These are ongoing expenditures which are now correctly reported here. This results in a seeming significant increase in serials expenditures and a corresponding decrease in expenditures for other materials.
21	Increase due to one-time money made available to purchase computer files in 2004-05.
24	Increase due to significant additional expenditures for hardware acquisitions and workstation life cycle replacements made in 2004-05.
26.a, 26.b	During 2004-05, several staff positions were reclassified as exempt. The bulk of their salaries were paid while they were classified as non-exempt. This change accounts for the changing numbers of support and professional staff, while at the same time showing little change in professional salary expenditures.
ILLINOIS, CHICAGO	
8, 12, 13, 21, 23.a, 25	Figures reflect Health Science Library only.
10	Maps librarian position was vacant for most of this time period, as a result this figure was unavailable.

QUESTION NUMBER	NOTES
17	Library of the Health Sciences information included in all Daley salary and wage figures. Breakdown of data for both locations is unavailable.
ILLINOIS, URBANA	
15.d	Includes: Postage/freight ($170,453); Handling ($66,170); Inter-Library Loan ($13,992); Copyright Fees ($31,914); Memberships ($66,129); Electronic Service Fees ($28,435); Online Non-Recoverable Reference Service ($24,209); Preservation ($90,383).
17.a	Includes $8,623,788 academic and $1,226,218 graduate assistants.
22	Increase due to increased efficiency at counting electronic serials, as well as the flipping of titles from print to electronic.
26.a	Includes 43 FTE graduate assistants.
33	Decrease attributable to decrease in renewals, which resulted from a newly enforced policy on number of renewals allowed.
34, 35	Excludes Law Library.
INDIANA	
	Includes the Law Library at Bloomington, the Archives of Traditional Music, and the Kinsey Institute, the Archives of African-American Music & Culture, Center for Disability Information & Referral, Gay Lesbian Bi-Sexual and Transgendered Student Support Services Office, and Workshop on Political Theory & Policy Analysis libraries. Excludes the regional campuses at Columbus, Ft. Wayne, Gary, Indianapolis (IUPUI), Kokomo, New Albany, Richmond, and South Bend. Volumes held June 30, 2004 revised to 7,131,059 to include collections from five new libraries: Archives of African-American Music & Culture, Center for Disability Information & Referral, Gay Lesbian Bi-Sexual and Transgendered Student Support Services Office, and Workshop on Political Theory & Policy Analysis.
20	Change in accounting practices eliminated double counting of some funds.
21	Improved accounting practices.
26.b	One unit was double-counted in 2003-04.
IOWA	
25	Postage was higher in 2003-04, thus charges could not be separated out correctly.
26.a	Change from 2003-04 due to retirements of over 8 Librarians.
38	Excludes 533 clinical and preclinical faculty with departmental appointments.
IOWA STATE	
	Includes the Veterinary Medical Library.
4.a	Decrease is a result of a correction in data reporting in 2004-05, i.e., full-text journals within indexing/abstracting products were excluded in 2004-05.
15.a	Increase due to addition of e-books.
15.c	Decrease is due to e-books being reported in (15a); also, over $100,000 less was spent on e-books in 2004-05 than in 2003-04.
15.d	Decrease reflects dollars that were reallocated to purchase monographs reported in (15a) and serials reported in (15b), plus a drop in available dollars in the Dean's discretionary budget line.
21	Decrease reflects a drop in available dollars in the Associate Dean for Collections & Technical Services' discretionary budget line to purchase one-time monographic items.
22	Decrease is primarily due to the payment of both 2003 and 2004 Science Direct annual subscriptions in 2003-04.

QUESTION NUMBER	NOTES
27	Closed the Economic and Sociology Reading Room July 2004.
JOHNS HOPKINS	
	Includes MSEL, Welch, SAIS, Friedheim, APL, and Montgomery County libraries.
KANSAS	
	Includes Regent Center Library, Edwards Campus (Overland Park, KS) and excludes University of Kansas School of Medicine, Farha Library, in Wichita, KS.
1.b.i	Used ARL formula of 5.2 documents pieces per volume for document serials
4.a	Figure reflects purchase of more electronic journals.
6	2003-04 figure revised to 3,584,065.
33	Includes Main and Law libraries only; Health Sciences reports U/A.
KENT STATE	
	All main campus libraries are included in volume counts, collection budget figures, and use statistics. All regional campus libraries are included in volume counts and use statistics, but not collection budget figures. Recent staffing changes may have resulted in reporting anomalies over the past several years.
10	Figure does not include map library.
15.b	Includes Paper ($1,185,536) and Electronic ($1,553,020).
21	Included in (15a).
KENTUCKY	
	Agricultural Information Center; Chemistry/Physics Library; Design Library; Education Library; Geological Sciences Library/Map Collection; Engineering Library; Fine Arts Library; International Documents/Storage (King Library); Mathematical Sciences Library; Special Collections & Digital Programs; William T. Young Library; Law Library and Medical Library statistics are reported separately. Volumes held June 30, 2004 revised to 3,212,669 to include e-books.
1.b.i	The Libraries added 3 large e-book collections during 2004-05.
11	Includes 1.5 million photographs from the Lexington Herald Leader project.
LAVAL	
	All figures are as of May 31, 2005. The Law and Health Libraries data are included with those of the Main Library.
15-25	Expenditures as reported in Canadian dollars: (15a) $1,932,926; (15b) $7,491,055; (15c) $152,203; (15d) $175,359; (15) $9,751,543; (16) $44,725; (17a) $3,820,078; (17b) $5,991,933; (17c) $253,730; (17) $10,065,741; (19) $603,272; (20) $20,465,281; (21) UA/NA; (22) UA/NA; (23a) UA/NA; (23b) UA/NA; (24) UA/NA; (25) UA/NA.
LIBRARY & ARCHIVES CANADA	
	All figures are as of March 31, 2005. In 2004 the former National Library of Canada (NLC) and the National Archives of Canada (NAC) merged to form the new Library and Archives Canada (LAC). Responses provided reflect all of LAC; 2003-04 figures reflected collections, budgets and activities of the NLC. Increases in some of those figures in 2004-05 reflect the merger of collections, budgets and activities of the new institution.
1.a	Figures are estimated.
9	Standard unit of measure for LAC Manuscripts and archives is linear meters; figure provided is converted to linear feet.
15-25	Expenditures as reported in Canadian dollars: (15a) $597,810; (15b) $868,867; (15c) $259,829; (15d)

QUESTION NUMBER	NOTES
	$23,973; (15) $1,750,479; (16) $6,679; (17a) $40,844,483; (17b) $21,310,912; (17c) $383,941; (17) $62,539,336; (19) $33,095,062; (20) $97,391,556; (21) $59,377; (22) $34,506; (23a) $42,878; (23b) UA/NA; (24) UA/NA; (25) $30,928.
25	Figure reflects the cost of a new microfilm scanner, maintenance for our other scanners, and upgrades to Relais software that we use to deliver articles to requesting libraries.
LIBRARY OF CONGRESS	
	All figures are as of September 30, 2005.
17.a	Figure reflects total salaries and wages, including that of support staff.
35	Includes only originals. This amount has decreased from 2003-04 because other institutions provided more facsimiles in lieu of originals.
LOUISIANA STATE	
	Volumes held June 30, 2004 revised to 3,361,209 to reflect electronic book records added: Netlibrary (46,000); Evans Digital Books (36,305); and Black Drama (1,195).
1.b.ii	Increase due to closing of Chemistry Library and elimination of duplicates.
2	Figure includes 20,965 paper and 45,461 e-books (netlibrary plus Evans Digital books, and Black drama).
4	Serials purchased (4a) and serials received but not purchased (4b) cannot be disaggregated.
11	Figure includes estimate of historic photographic collection, plus the aerial photographs.
31, 32	Excludes Health Science Library.
LOUISVILLE	
	Includes Main Library, Health Sciences Library, Art Library, Music Library, University Archives, and Law Library. Volumes held June 30, 2004 revised to 1,950,620.
29-31, 33-35	Excludes Law Library.
McGILL	
	All figures are as of April 30, 2005 and include the following libraries: Birks Reading Room (Religious Studies); Blackader Lautermann (Art & Architecture); Blacker-Wood (Biology); Education; Islamic Studies; Management; Macdonald Campus (Agriculture); Music, Humanities & Social Sciences; Life Sciences & Osler; Rosenthall Mathematics & Statistics; Hitschfeld Geographic Information; Science & Engineering; Law.
15-25	Expenditures as reported in Canadian dollars: (15a) $3,444,702; (15b) $8,145,878; (15c) $1,541,366; (15d) $1,085,545; (15) $14,217,491; (16) $212,407; (17a) $5,196,316; (17b) $5,909,602; (17c) $1,033,912; (17) $12,139,830; (19) $4,648,687; (20) $31,218,415; (21) $827,473; (22) $6,476,388; (23a) $227,491; (23b) $0; (24) $1,014,385; (25) $248,596.
McMASTER	
	All figures are as of April 30, 2005 and include Mills Memorial Library, H.G. Thode Library, Innis Library, and the Health Sciences Library.
1.b	Decrease due to volumes withdrawn in the medical library.
2	Increase due to 25,865 e-books not included in 2003-04.
4.a	Increase due to participation in more consortial packages for e-journal suites.
4.b	Increase due to more open-access journals.
15-25	Expenditures as reported in Canadian dollars: (15a) $1,522,233; (15b) $5,389,049; (15c) $390,362; (15d) $177,764; (15) $7,479,408; (16) $190,374; (17a) $2,259,189; (17b) $3,876,568; (17c) $363,321; (17) $6,499,078; (19) $893,379; (20) $15,062,239; (21) $154,813; (22) $3,757,904; (23a) $82,909; (23b) $0; (24)

QUESTION NUMBER	NOTES
	$252,410; (25) $101,106.
15.c	Increase due to maps, electronic backfiles (JSTOR), and memberships (CRKN and DLI) being included for the first time.
15.d	Increase due to brokerage, shipping, and handling expenditures that were counted under (15a) in 2003-04.
25	Decrease due to a consortial interlending system (RACER).
MANITOBA	
	All figures are as of March 31, 2005 and include the William R. Newman (Agriculture), Architecture/Fine Arts, D.S.Woods (Education), Elizabeth Dafoe, Donald W. Craik (Engineering), Neil John Maclean (Health Sciences), St. Boniface Hospital, Victoria General Hospital, Grace General Hospital, Concordia General Hospital, Seven Oaks General Hosptial, E.K. Williams (Law), Albert D. Cohen (Management), Eckhardt-Grammattee (Music), St. John's College, Fr. Harold Drake and Sciences and Technology Libraries. Law and Health Sciences data cannot be disaggregated from the rest of the ARL Statistics. Volumes held June 30, 2004 revised to 2,125,287.
1.b	Includes e-books.
4, 4.a	Serials from aggregated packages excluded from 2003-04, but included in 2004-05.
15-25	Expenditures as reported in Canadian dollars: (15a) $1,755,902; (15b) $5,299,257; (15c) $251,947; (15d) $645,739; (15) $7,952,845; (16) $115,978; (17a) $4,500,640; (17b) $5,065,751; (17c) $708,600; (17) $10,274,991; (19) $1,646,623; (20) $19,990,437; (21) $519,750; (22) $3,077,170; (23a) UA/NA; (23b) UA/NA; (24) $331,235; (25) $232,868.
15.b	2003-04 figure included some expenditures on print serials that were double counted.
32	2003-04 figure included reserves; 2004-05 figure does not.
33	2003-04 figure included internal circulation and should be revised to 552,383.
MARYLAND	
	Volumes held June 30, 2004 revised to 3,182,973 to include 100,000 e-books as part of the EEBO collection in 2003.
2	Increase is a one-time jump reflecting a change in the way monographs data are counted.
26	Total FTE includes cost recovery positions, but not special projects or grants.
34, 35	2003-04 figures revised to (34) 16,297; (35) 28,804.
MASSACHUSETTS	
	Includes DuBois Library, Integrated Science and Engineering Library, and the Music Reserve Lab. Volumes held June 30, 2004 revised to 3,169,283.
17.c	Includes $119,855 in federal work-study funds.
36	Excludes 43 Ed.D. degrees awarded.
37	Excludes 1 Ed.D progra, and 1 Au.D. (professional degree in Audiology) program.
MIT	
	Includes all branches in the MIT Libraries system.
15.d	Figure includes postage and service charges for serials.
MIAMI	
	All figures are as of May 31, 2005, except for enrollment data (lines 39 - 42), and include Otto G. Richter Library (central library) and its branches (architecture, business, math, music, and remote storage facility) and the law, marine, and medical libraries at the University of Miami.

QUESTION NUMBER	NOTES
4.a, 4.b	Data are not available for central library but are available for law, marine and medical libraries. Medical library includes print serials only. E-serials accessed only through aggregators are not included.
8, 13	Includes computer files in medical library.
15.b	Medical library figure includes only print titles.
15.c	Includes medical library expenditures for electronic serials.
17.a	Excludes salaries of the university librarian and directors of the law and medical libraries; includes salaries of non-librarian professionals in central library.
17.c	Excludes wages for student assistants subsidized by the School of Business Administration.
21-25	Marine library expenditures for electronic materials are not available.
24	2003-04 figure revised to $372,543 as IT hardware purchases of $81,480 for the central library's information commons were omitted. 2004-05 figure includes $339,959 for computer equipment purchased for the Weeks Music Library and Technology Center that opened in January 2005 and $185,795 for hardware purchased for the central library's information commons.
29, 30	Business library was double-counted in 2003-04; figures revised to (29) 3,487; (30) 18,324. Medical library count includes only group presentations in 2004-05 but included individual instruction sessions as well in 2003-04.
31	Reference transactions for marine and architecture libraries are unavailable.
31.a	Reference transaction totals are based on samples in law and medical libraries. For central library reference transactions are enumerations for all service points except special collections, which is based on sampling.
MICHIGAN	
	Includes the Bentley Historical Library, William L. Clements Library, Law Library, Kresge Business Administration Library, and the Univeristy Library. Excludes the libraries at University of Michigan-Dearborn and University of Michigan-Flint.
MICHIGAN STATE	
	Includes Main Library, Law Library, Biomedical and Physical Sciences Library, Business Library, Engineering Library, Geology Library, International Center Library, Math Library, Planning and Design Library, and the Veterinary Medical Center Library. Volumes held June 30, 2004 revised to 4,670,003 to include e-books (NetLibrary and government documents).
4, 15.b	Main Library includes monographic series, whose counts and expenditures cannot be disaggregated.
15.c	The Main Library expenditures are included in monograph expenditures (15a) and current serials expenditures (15b). Figure includes the Law Library only.
15.d	Main Library includes expenditures for ILL Reserve $85,780; CRL Reserve $43,215.
22	Increase due to the conversion of large numbers of journals from print-based pricing to a model where the online version is the basis of the subscription.
23.a	Main Library figure reflects significant savings achieved from 2003-04 due to reductions in telecomm and searching costs.
23.b	The MSU Libraries have access to a large number of electronic resources from several vendors through the Michigan eLibrary. These resources are provided to all residents of the State of Michigan through academic, school, and public libraries. Thus, meaningful figures on MSU's financial share of the total expenditure are unavailable.
MINNESOTA	
	Excludes the University of Minnesota coordinate campus libraries: University of Minnesota

QUESTION NUMBER	NOTES
	Crookston, University of Minnesota Duluth, and University of Minnesota Morris. Volumes held June 30, 2004 revised to 6,119,734 to include revised collection counts and government documents added to OPAC.
1.b	Includes e-books.
1.b.ii	Reflects duplicate withdrawal project.
2	Count reflects monograph titles ordered, not volumes.
4	Increase reflects more comprehensive counts of electronic serials and standing orders.
4.a	Increase due to the correction of an error in the method of counting electronic serials and standing orders in the library system.
7	Adjusted for 5,358 volumes added to OPAC.
11	Breakdown for Audiovisual materials not available. Total AV materials count, including Health Sciences and Law, is 857,110.
15.c	Increase reflects expenditures on backfiles of electronic serials.
19	Overall increase largely due to change in University budgeting for support services for library systems.
21	Decrease reflects a decline in the purchase of computer files as defined, plus a more accurate count of expenditures on this class of materials.
24	Increase largely due to change in University budgeting for some support services for library systems.
31.a	Reference transactions are mainly counts; some units' figures are estimated.
38	Full-time primarily instruction non-medical faculty reported in 2004-05 EAP survey.
41	Includes all post-baccalaureate (graduate/professional).
MISSOURI	
	Includes Vet-med, Math, Geology, Newspaper, Journalism, Engineering, Archives, and Health Sciences libraries. Health Sciences data cannot be disaggregated from the rest of the ARL Statistics.
4.a	Figure reflects a more accurate count of TDNet serials.
15.b	Includes $773,414 from University Wide Library System office for E-Serials/databases.
17.a	Includes $194,267 of S&W from University Wide Library Systems office allocated to MU operations.
19	Includes $143,546 for MERLIN catalog system from Library Systems Office, and $681,485 in capital projects from gift funds/salary savings.
22	Includes $773,414 from Library Systems Office from University wide funds for E-Serials/Databases.
38	Decrease due to not counting any faculty that are Research Faculty, as opposed to "teaching faculty." Only IPEDS SA Fall Instructional Faculty counts were used.
MONTREAL	
	All figures are as of May 31, 2005 and include: Aménagement, Bibliothéconomie et sciences de l'information, Botanique, Chimie, Didacthèque, ÉPC-Biologie, Géographie, Kinésiologie, Lettres et sciences humaines, Livres rares et collections spéciales, Mathématiques et informatique, Médecine vétérinaire, Musique, Optométrie, Physique, Dépôt central.
1.b	Decrease due to weeding at Ecole Polytechnique Library, which has moved into a new building.
4.a	Increase due to a revised figure at HEC Montreal.
15-25	Expenditures as reported in Canadian dollars: (15a) $2,991,897; (15b) $8,576,085; (15c) $105,744;

QUESTION NUMBER	NOTES
	(15d) $220,720; (15) $11,894,446; (16) $271,302; (17a) $7,079,321; (17b) $10,109,730; (17c) $55,156; (17) $17,244,207; (19) $1,461,090; (20) $30,871,045; (21) $294,467; (22) $4,746,921; (23a) $141,323; (23b) $0; (24) $860,675; (25) $149,410.
27	Figure revised in order to follow the instructions more closely.

NATIONAL AGRICULTURAL LIBRARY

	All figures are as of September 30, 2005. Volumes held September 30, 2004 revised to 2,396,472.
4.a	Figure accurately reflects NAL purchasing during 2004-05.
15.a	Decrease due to budget limitations; NAL chose to expend fewer funds on monographs in 2004-05.
27	Total includes library reference and information center service points, plus special collections, AgNIC, Current Awareness Literature Service, and invasive species information.
29	Number of library presentations includes demonstrations during exhibits.
35	The large decrease is attributable to the increasing availability of articles available electronically in full text via the library's DigiTop digital desktop library, which has replaced the need for document delivery in many cases.

NATIONAL LIBRARY OF MEDICINE

	All figures are as of September 30, 2005.

NEBRASKA

	Volumes held June 30, 2004 revised to 2,669,484 to reflect inclusion of Early English Books Online (EEBO).
4	Increase due to the purchase of more electronic serials over the past few years.
7-9	Figures unavailable for Law Library.

NEW MEXICO

	Includes Centennial Science & Engineering Library, Fine Arts Library, Parish Memorial Library, Zimmerman Library, Bunting Visual Resources Library, Tireman Library, and Native American Library. Excludes Gallup Branch Campus Library, Los Alamos Branch Campus Library, Taos Branch Campus Library, Valencia Branch Campus Library.
12	Decrease due to deaccessioning.
13	2003-04 data incorrect.
15.a	Decrease due to bond funds not available this year and budgetary constraints.
15.b	Increase due to revision of coding for identification of electronic serials.
15.c	Increase due to purchase of digital backfiles with bond funds.
15.d	Decrease due to changes in vendor billing cycles.
17.c	Decrease due to budgetary constraints.
25	Increase due to new consortium, added delivery charges, and an increase in loaned items.

NEW YORK

	All figures are as of August 31, 2005, and include Bobst, IFA, REI & Courant Libraries. Volumes held revised to 3,871,739 due to a correction of 150,000 e-book packages in 2003-04 and a further addition of 64,854 e-book packages in 2004-05.
23.a, 23.b	Figures include Law, Medical, and Dental libraries only.

NEW YORK PUBLIC

4.b	Acquisitions Division undertook assessment of serials received, resulting in more accurate count.

QUESTION NUMBER	NOTES
34	Inter-library document delivery functions were consolidated in ILL office out of corporate research division, resulting in more accurate count.

NEW YORK STATE

All figures are as of March 31, 2005.

NORTH CAROLINA

Includes University Library (Academic Affairs), Health Sciences, Law, and Institute for Marine Sciences libraries. Includes the Sonja Hanes Stone Center for Black Culture and History, which opened fall 2004 (Most monographs were purchased for the Stone Center library in 2003-04 and were reported as purchased in that year, but were added to the collection in 2004-05).

1.a, 1.b.i	Includes government documents.
15.d	Miscellaneous library material expenditures include payments for CRL membership.
20	Includes $6,144,864 in grants, trust, and other special funds; excludes $234,621 in expenditures for the Triangle Research Libraries Network from funds contributed by Duke, North Carolina State, and North Carolina Central Universities.
26	Includes 2.76 professionals on special grant funds; .37FTE on special projects. Excludes 2.0 FTE paid from funds contributed by Duke, North Carolina State, and North Carolina Central universities for support of the Triangle Research Libraries Network.
32	Data unavailable for Main Library.

NORTH CAROLINA STATE

Includes Design, Natural Resources, Textiles, and Veterinary Medical libraries.

1	Includes 171,460 e-books.
2	Includes some monographic standing order items paid from serials funds.
12	The growth rate for 2004-05 reflects the normal range of purchasing activity for NCSU, augmented by 301 CDs from the Compact Disc Minimum Advertised Price Antitrust Litigation Settlement.
15.a	Some monographic titles are paid from serials funds included in (15b).
15.d	Includes bibliographic utilities ($168,823); document delivery ($139,934); and memberships ($206,971).
21, 23.a	Decrease due to unusually high figures in 2003-04.

NORTHWESTERN

All figures are as of August 31, 2005 and include the Science and Engineering, Math, and Geology libraries, and the Law, Health Sciences and Schaffner (business) libraries on the Chicago campus.

1-3	Includes government documents.
6, 8, 9, 11	Excludes Health Sciences Library.
9	Excludes Law Library.
22	OCLC credits removed from total.
39-42	2003-04 figures were reported incorrectly.

NOTRE DAME

1.a	Includes 3,943 e-books.
2	Includes 431 e-books.
15.d	Miscellaneous materials expenditures consist primarily of memberships for the purpose of obtaining/accessing publications and literature searching.

QUESTION NUMBER	NOTES
21	Increase due to several large one-time database purchases in 2004-05.
22	Electronic indexes and reference tools total $746,374. Electronic full text periodicals total $2,233,715.
24	Increase due to a significant technology purchase being delayed from 2003-04 to 2004-05.
29-31	Figures unavailable for Law Library.
OHIO	
	Includes Athens (main), Chillicothe, Eastern, Lancaster, Southern, and Zanesville libraries.
OHIO STATE	
	Includes the following libraries: Architecture; Biological Sciences/Pharmacy; Business; Cartoon Research; Children's Hospital; Education, Human Ecology, Psychology, and Social Work; Fine Arts; Food, Agricultural and Environmental Sciences; Geology; Health Sciences; Journalism; Law; Music/Dance; Science & Engineering; Theatre Research Institute; University Archives; and Veterinary Medicine. Excludes Agricultural Technical Institute Library, Ohio Agricultural Research and Development Center and Regional Campus Libraries at Lima, Mansfield, Marion, and Newark.
1.b.ii	System improvements made during the year eliminated a number of duplicate records, which resulted in a positive number of volumes "withdrawn."
4.a	Decrease due to a significant cancellation project in the Main Library, and also some titles being counted as "purchased" that should have been counted as "not purchased."
4.b	Includes 3,865 titles incorrectly counted among paid titles in prior years.
13	Audio, Film and Video are included as a single category in our catalog.
15.b	Figure reflects the cancellation project, as well as the absence of a $2 million prepayment that had been made in prior years.
15.c, 15.d	In earlier years these figures had not been reported separately, but had been counted as monograph expenditures for one time purchases or as a serial expenditure for continuations.
17	2003-04 figure revised to $13,077,736.
17.c	2003-04 figure revised to $1,483,565; the Work Study figure had not been included for the Main Library.
32	2003-04 figure revised to 598,717.
34, 35	Includes materials loaned and borrowed via the OhioLINK statewide system.
OKLAHOMA	
4.a	Added several new major databases containing multiple serials.
4.b	Increase due to a change in the reporting process for serials received but not purchased.
7	A number of commercially published volumes were purchased and several gifts were received in 2004-05.
15.a, 19, 22, 23.a	Increase due to an increase in the library budget.
21	Decrease due to a change in reporting method from 2003-04.
24	Increase due to additional computer hardware purchases.
36-42	Previous years did not include medical campus.
OKLAHOMA STATE	
	Volumes held June 30, 2004 revised to 2,551,684 due to the over-reporting of 20,360 volumes from a campus library.

QUESTION NUMBER	NOTES
4	Print and e-journals received less aggregator packages such as LexisNexis, EBSCO, Proquest, etc.
OREGON	
	Includes Knight Library, Architecture and Allied Arts Library, Science Library, Math Library, Portland Architecture Library, and Rippy Library (Oregon Institute of Marine Biology).
4.a	Increase due to a change in the reporting method for print plus electronic serials.
9	Figure is in linear feet.
15.c	Expenditures included in monographs (15a) and serials (15b).
19	Decrease due to an unusually high figure in 2003-04 in support of campus-wide Educational Technology projects.
21	Decrease due to a three-year payment for the Web-of-Science backfile coming to an end.
22	Increase due to being better able to track expenditures for electronic subscriptions.
25	Increase due to a subsidized pay-per-view service that was instituted.
32, 33	Past data included ILL transactions and internal circulation.
34, 35	Law Library data cannot be disaggregated.
38, 39, 41	2003-04 figures revised to: (38) 812; (39) 16,676; (41) 2,885.
PENNSYLVANIA	
	Volumes held June 30, 2004 revised to 5,660,177 to include titles from several large e-book sets which are cataloged separately, and for which Penn owns tape backups and/or long-term access agreements with publishers.
31	Health Science Library data are unavailable.
34, 35	Includes patron direct borrowing through regional and Ivies consortia.
PENNSYLVANIA STATE	
	Includes Penn State Beaver, Penn State Delaware, Penn State DuBois, Penn State Fayette, Penn State Hazleton, Penn State Lehigh Valley, Penn State McKeesport, Penn State Mont Alto, Penn State New Kensington, Penn State Schuylkill, Penn State Shenango, Penn State Wilkes-Barre, Penn State Worthington Scranton, Penn State York, Penn State Abington, Penn State Altoona, Penn State Behrend, Penn State Berks, Penn State Great Valley, and Penn State Harrisburg.
2	Includes 195 e-books, and 39 Gale Virtual Reference Library titles.
31	Extrapolation used: 41.2987.
PITTSBURGH	
	Includes the University Library System, the Health Sciences Library, and the Law Library.
6	Increase due to a large purchase of microforms.
8	Statistics gathering has been streamlined, and computer files were counted differently. Future figures will be consistent with this process.
PRINCETON	
4.a	Count represents an increase in electronic holdings and includes a fuller, more accurate count of aggregator titles made possible by the Library's new Electronic Resource Management system.
7	Count reflects an adjusted base figure more accurately differentiating uncataloged vs. cataloged print documents made possible by improved management data reporting of the Library's current Integrated Library System.
11	Increase represents the one-time transfer to the Library of the University's administrative photo archives.

QUESTION NUMBER	NOTES
15.d	Expenses for Document Delivery, previously recorded as Other Library Materials (15c), are now accounted for here.
23.a	2003-04 figure included expenditures for Marcive, Palinet, and Dialog that are not included here.
24	Includes one-time set up cost for Digital Imaging Studio.
25	2003-04 figure revised to $165,440 due to under reporting by $71,670.
PURDUE	
	Excludes branch campuses at Purdue North Central in Westville, Purdue Calumet in Hammond, and Purdue Fort Wayne in Fort Wayne.
1-16	Includes cataloged government documents.
2	Represents number of titles rather than volumes. Excludes materials purchased on sponsored research programs.
3	Basis of volume count is physical for print materials and bibliographic for electronic.
4	Excludes some numbered monographic series.
4.b	Includes cataloged depository, exchange, and gift serials.
15.c	Represents bibliographic and full-text databases for which expenditures can be tracked separately. Expenditures for other databases and for materials such as maps and audiovisuals are included in those for monographs (15a) and serials (15b), and cannot be disaggregated.
22	Includes expenditures for bibliographic and fulltext databases in (15c), plus electronic journals.
25	Includes $130,761 of expenditures recovered by fee-based service.
29	29-30 Excludes users of online, interactive tutorial, "CORE" - approximately 2,100 account holders, plus unregistered users made 40,968 visits in 2004-05.
31	Includes 2,601 reference transactions via digital reference service.
QUEEN'S	
	All figures are as of April 30, 2005, and include: Stauffer Library (Humanities and Social Sciences), Douglas Library (Engineering/Science), WD Jordan Library (Special Collections & Music), Bracken Library (Health Sciences), Law Library, Education Library.
1.b.i	2003-04 figure was unusually high due to a backlog cleanup project.
15-25	Expenditures as reported in Canadian dollars: (15a) $1,399,787; (15b) $6,425,128; (15c) $99,102; (15d) $533,882; (15) $8,457,899; (16) $117,508; (17a) $2,333,401; (17b) $4,448,190; (17c) $321,331; (17) $7,102,922; (19) $2,258,385; (20) $17,936,714; (21) $0; (22) $4,329,595; (23a) $248,650; (23b) $0; (24) $294,846; (25) $45,263.
23.a	Increase due to an increase in contract cataloging.
25	Decrease due to offsetting revenues' increase, as Queen's became a net lender of material.
34	Data for Law Library unavailable; increase due to Queen's becoming a net lender of material.
35	Data for Law Library were unavailable.
36-42	Figures are as of November 2004.
38	Figure as of October 2004.
RICE	
	Includes Fondren Library, Brown Fine Arts Library, and Business Information Center. Volumes held June 30, 2004 revised to 2,394,131 and includes 23,466 Texshare e-books, 23,872 government document e-books, and 630 Netlibrary e-books.
2	Includes 2,562 government document e-books.

QUESTION NUMBER	NOTES
9	Manuscripts and archives reported in cubic feet.
15.d	Includes memberships ($163,184), ILL ($33,446), in-house binding supplies ($13,060), and bibliographic utilities ($116,445).
17.b	Decrease due to changes in positions and organizational units following retirement incentive program, as well as elimination of several positions previously supported through special funds.
21	Increase due in part to purchase of several large digital collections.
22	Decrease due to more thorough exclusion of funds spent on aggregator journals excluded from subscription counts.
24	Increase due to special funding not available in 2003-04.
ROCHESTER	
	The Main library includes statistics from the River campus Libraries as well as statistics from the Sibley Library of the Eastman School of Music. The Health Sciences Library statistics are from the Edward G Miner Library of the School of Medicine and Dentistry. Volumes held June 30, 2004 revised to 3,265,087 to include 148,325 volumes acquired from Swasey Library, Colgate Rochester Divinity School.
3	For 2,480 volumes the basis of the count was physical.
4.a, 4.b	Main library reports (4a) 21,394; (4b) 932; Health Science Library data unavailable.
7	No retroactive government document records were loaded in 2003-04, so this figure has not changed.
9, 24	Data not available from Sibley Library.
15.c	Includes scores, recordings, microforms, and backfiles.
18	Fringe benefits are included for all but $636,346 of salaries and wages.
29	148 of total based on sampling.
30	2,900 of total based on sampling.
32	Figure excludes recordings.
33	Figure excludes recordings and no longer includes self-renewal statistics.
RUTGERS	
	Includes two Law libraries, services to a non-Rutgers medical school and campuses in Camden, Newark and New Brunswick.
4.a	Increase due to purchase of e-journal packages and cataloging of individual titles within the packages.
10-13	Audiovisual Materials figures cannot be disaggregated. Total audiovisual is 144,030.
17, 26	Excludes staff hired for special projects and grants.
23.a	Increase due to payments made over multiple years and moving cataloging from RLG to OCLC.
SASKATCHEWAN	
	All figures are as of April 30, 2005 and include the Main Library plus Education, Engineering, Natural Sciences and Veterinary Medicine branch libraries. Excludes Theological College libraries on campus that are not part of the library system.
1.b	Decrease due to weeding project.
4	Changes in database have enabled more accurate reporting of electronic materials.
15-25	Expenditures as reported in Canadian dollars: (15a) $2,122,006; (15b) $5,659,104; (15c) $0; (15d) $195,186; (15) $7,976,296; (16) $93,817; (17a) $3,096,030; (17b) $3,114,477; (17c) $337,072; (17)

QUESTION NUMBER	NOTES
	$6,547,579; (19) $1,034,790; (20) $15,652,482; (21) $11,065; (22) $3,234,567; (23a) $65,412; (23b) $0; (24) $422,711; (25) $125,617.
15.d	Increase due to a payment of $65,000 for ILL from 2003-04 not paid until fiscal year 2004-05.
17.a	Computer programmers, technicians and budget officers included in professional staff.
20	Includes funding from grants and special funds.
22	Includes $217,960 in provincial funding for resources for the Saskatchewan Health Information Resources Partnership; unable to determine the exact portion of this $217,960 that would be specific to the U of S.
38	Includes professors, associate professors, assistant professors, lecturers and instructors.
39	Includes 99 certificate/diploma students.
40	Includes 559 certificate/diploma students.
41	Includes 221 post-graduate clinical students.
SMITHSONIAN	
15.c, 15.d	In 2004-05 moved membership dues from Other Materials (15c) to Miscellaneous Materials (15d).
23.a	2003-04 figure in (23b) should have been recorded in (23a).
SOUTH CAROLINA	
	Includes Main, Law and Medicine, Music, Math, and South Caroliniana Libraries; excludes Aiken, Beaufort, Spartanburg (Upstate), Sumter, Union, Lancaster, and Salkehatchie branch campus libraries.
SOUTHERN CALIFORNIA	
	Law and Health Sciences libraries were inadvertently excluded from 2003-04 figures; as a result, volumes held June 30, 2004 revised to 3,865,914.
21	Includes Health Sciences Library only.
26.a	Reassessment of previously reported data.
SOUTHERN ILLINOIS	
	Volumes held June 30, 2004 revised to 2,965,957 to include onetime downloads for EEBO (74,142) and LiOn (15,491) in 2003-04, and 36,000 e-books from Early American Imprints, Series II.
4.a	Increase reflects improved counting methods and additional titles in packages of electronic journals.
11	Decrease reflects the weeding of materials in the Curriculum Materials Center.
15.c	Includes maps, serials backfiles, spoken word files, manuscripts, films/videos, microforms, music, and software.
15.d	Includes bibliographic utilities.
21	Includes archive capital fees for electronic collections and A/I databases, periodical backfiles, and electronic books collections maintenance and access fees.
26.a, 26.c	Grant FTE are (26a) 2; (26c) 1.
SUNY-ALBANY	
	Includes Gov. Thomas E. Dewey Graduate Library of Public Affairs and Policy Science Library.
4.a	Figure reflects a title count including electronic journals from all sources; those accessed through indexing and abstracting products cannot be distinguished from those provided by other means.
20	$266,307 of the total represents expenditures from private giving to the Libraries.

QUESTION NUMBER	NOTES
SUNY-BUFFALO	
36	2003-04 figure revised to 299.
SUNY-STONY BROOK	
4.b	Increase reflects the number of electronic journals not purchased, as reported on the 2004-05 *ARL Supplementary Statistics*.
15	Includes $142,890 in research funds and grants.
16	Includes $4,612 in funds from NYS preservation grant.
17	Figure includes librarians and other professionals.
38	Figure includes Health Sciences Center.
SYRACUSE	
	Includes Science and Technology, Physics, Geology, and Law libraries.
15.c, 23.a, 23.b	Includes Law Library only.
TEMPLE	
	Main campus libraries include Paley, Engineering & Architecture, Charles L. Blockson Afro-American Collection, Chemistry, Mathematical Sciences, Physics, Zahn, and Law. Branch campus libraries include Ambler, Tyler School of Art, Harrisburg, and the Health Sciences Center. Volumes held June 30, 2004 revised to 2,449,753 to include 551 records added for a portion of the Arabic collection.
1.b.i	Decrease due to not binding print versions of ScienceDirect journals, as well as other titles for which we receive print and electronic versions. Decrease also due to a high number of bound journal volumes in 2003-04, which resulted from a temporary binding halt.
2	2003-04 figure revised to 31,183 due to incorrect reporting by the Main Library.
4	Increases reflect improved method of counting e-journals.
7	Government documents received and cataloged during 2004-05 are counted under (1bi). There was no retrospective conversion done on government documents during 2004-05.
17.a	Half of development officer salary is paid by central development office.
17.b	Includes expenditure for staffing security desk at main library, funded by University budget.
17.c	Increased number of student assistants and raised wages.
21	Major purchases of backfiles and e-books, including Royal Society of Chemistry, Nature, Web of Sciences, Alexander Street Press, and Readex.
23.a	Figures more precisely exclude expenditures for user access to consortial databases.
23.b	Estimated benefit share of state-funded databases (Access PA).
26.a	Decrease reflects narrower interpretation of "professional."
26.b	Increase reflects narrower interpretation of "professional" in (26a). Includes security guard for main library, funded by University budget.
36	A number of PhDs that were not completed at the conclusion of 2003-04 were completed early in 2004-05, thus increasing the overall total.
TENNESSEE	
	Includes the University of Tennesse Knoxville, Law Library, and Social Work Branch in Nashville; the Preston Medical Center Library; the UT Space Institute in Tullahoma, TN and the University of Tennessee at Memphis.
4.a	Decrease due to a revised interpretation for counting serials purchased.

QUESTION NUMBER	NOTES
24	Increase due to a significant upgrade of staff hardware.
29, 30, 32	Data unavailable for Law Library.
TEXAS	
	All figures are as of August 31, 2005 and include the University of Texas Libraries, the Center for American History, the Tarlton Law Library, and the Humanities Research Center. Volumes held June 30, 2004 revised to 8,814,148 to include e-books, including netLibrary, Eighteenth Century Collections Online, Early English Books Online, and Gale Virtual Reference Library.
1.b	Decrease due to withdrawals from both Law and UT Libraries.
2	Includes 43,921 monographic volumes purchased by the University Libraries, 188 by the Center for American History, 3,301 by the Tarlton Law Library, and 2,306 by the Humanities Research Center.
4.a	Includes 20,650 serials purchased by the University Libraries (13,933 print and 6,717 electronic), 108 by the Center for American History, 5,496 by the Tarlton Law Library, and 179 by the Humanities Research Center.
4.b	Includes 17,364 serials received but not purchased by the University Libraries (9,045 print and 8,319 electronic), 198 by the Center for American History, 2,826 by the Tarlton Law Library, and 36 by the Humanities Research Center.
15	Includes expenditures of $10,590,602 by the University Libraries, $49,109 by the Center for American History, $1,173,643 by the Tarlton Law Library, and $1,336,040 by the Humanities Research Center.
15.a	Includes expenditures of $3,397,307 by the University Libraries, $4,850 by the Center for American History, $224,981 by the Tarlton Law Library, and $57,466 by the Humanities Research Center.
15.b	Includes expenditures of $6,946,054 by the University Libraries, $5,862 by the Center for American History, $891,570 by the Tarlton Law Library, and $18,681 by the Humanities Research Center.
15.c	Includes expenditures of $246,650 by the University Libraries, $38,397 by the Center for American History, $57,092 by the Tarlton Law Library, and $1,259,893 by the Humanities Research Center.
15.d	Includes expenditures of $591 by the University Libraries.
17, 19, 26	Salaries, wages and other operating expenditures reflect increased expenditures for staff and content development for UTopia, an outreach service to the citizens of the State of Texas. Total staff (26) includes 1.5 FTE staff hired on grant funds.
17.c	Includes federal work-study funds.
30	Includes 34,614 users of online tutorial.
36	Does not include Ed.D degrees.
TEXAS A&M	
	Volumes held August 31, 2004 revised to 3,331,124 to reflect the addition of 20,284 records for electronic packages.
4.a	Increase is a result of changing software that allows for a more accurate count of serials offered electronically.
7	Figures unavailable for branch libraries.
15.b	Databases were counted under Other Library Materials (15c) in 2003-04. The change or clarification of the term moved the expenditure to (15b) in 2004-05.
15.d	Bibliographic utilities are included as Miscellaneous (15d) for 2004-05; were previously reported erroneously under Other Operating Expenditures (19).
22	Increase due to some materials being reclassified from databases to serials.
23a, 23b	2003-04 figures for these items should be reversed.

QUESTION NUMBER	NOTES
24	Increase due to the purchase of software and scanning equipment for 2004-05.
26.a	2003-04 figure revised to 128 to include graduate assistants inadvertently excluded.
26.b	2003-04 figure revised to 140.

TEXAS TECH

All figures are as of August 31, 2005 and include the Architecture Library, Museum Library, Geosciences Library, Education Library, International Cultural Center Library, Vietnam Archive Library, and the Southwest Collections/Special Collections Library. Volumes held June 30, 2004 revised to 1,930,563 to include books from the Education Library not previously included. The formula of "10 traditional volumes per foot" was used to calculate government documents at the University Library.

TORONTO

All figures are as of April 30, 2005 and include the following libraries: Architecture, Astronomy, Business Information Centre, Criminology, Dentistry, Family and Community Medicine, Industrial Relation, Information Studies Forum, Innis College, Knox College, Law, Massey College, Mounst Sinai Hospital Library, Music, New College, Physics, Royal Ontario Museeum, Regis College, St. Augustine Seminary, St. Michael College, Trinity College, University College, U of T at Mississauga, U of T at Scarborough, University Schools, Victoria University, and Zoologu. Volumes held June 30, 2004 revised to 10,101,550 due to recount of materials.

1.b.ii	Numerous materials were weeded in 2004-05.
4.a	Includes electronic serials.
8	Increase due to the acquisition of a large collection.
15-25	Expenditures as reported in Canadian dollars: (15a) $8,660,326; (15b) $11,805,665; (15c) $4,298,749; (15d) $0; (15) $24,764,740; (16) $445,951; (17a) $12,777,836; (17b) $16,634,316; (17c) $4,056,892; (17) $33,469,044; (19) $7,678,993; (20) $66,358,728; (21) $1,067,114; (22) $2,874,228; (23a) $426,687; (23b) $0; (24) $4,129,180; (25) $187,989.
15.d	Library materials from trust funds are included in questions (15a) through (15c).
25	Includes Main Library only.
34	An increase in ILL/Document Delivery activities was due to (a) the re-opening of Resource Sharing Section at U of T at Scarborough Library after renovation was done; and (b) the introduction of an automated ILL Request system.
40	Decrease is due to the fact that effective fall 2004, the University no longer reports OISE/UT part-time Additional Qualification students. This has resulted in a significant downward adjustment to the multi-year time series.

TULANE

Due to damage inflicted by Hurricane Katrina, Tulane University was unable to submit data. The 2003-04 data appear instead, with the library's permission. These data include the Architecture, Business, Law, and Women's Center libraries, as well as the Medical branch campus library.

UTAH

1.b	Decrease due to Main Library withdrawing 13,519 duplicates.
4	Law Library data unavailable.
7	2003-04 figure revised to 879,319.
8	Main library withdrew 1,882 volumes.
9	Decrease due to 13,867 Architectural Drawings withdrawn.
15.c	Includes Health Sciences Library only.

QUESTION NUMBER	NOTES
15.d, 21	Includes Law and Health Sciences libraries only.
VANDERBILT	
	Includes Biomedical, Central, Divinity, Law, Management, Music, Peabody, Science & Engineering, and Special Collections libraries. Volumes held June 30, 2004 revised to 2,991,563 to reflect addition of a 22,973 volume e-book collection in Law Library, and 4,375 EEBO records.
1.b.i	Includes 9,111 e-books in Eighteenth Century collections online added in 2004-05 by Law Library.
2	E-book collection purchased with single order; emphasis on journal purchasing.
4	Includes open access and other free e-journals not counted before; KnowledgeBase makes more accurate count possible.
VIRGINIA	
	Excludes the UVa College at Wise Library.
VIRGINIA TECH	
	Includes Art & Architecture and Veterinary Medicine libraries.
WASHINGTON	
	Includes the Law and Medical libraries, except where noted. Medical library includes Health Sciences Library, Social Work Library, Harborview Medical Center, and Regional Medical Library (National Network/Libraries of Medicine. Excludes Bothell and Tacoma campuses.
1.b.i	Decrease due to 2003-04 figure counting e-books for the first time.
2, 21, 31	Figures unavailable for Law Library.
4	Increase due to additional e-serial subscriptions.
24, 25	Figures available only for Law Library and Medical Library.
33	Previous years incorrectly included in-library use.
34, 35	Includes books loaned to and borrowed from Cascade-Orbis consortium members.
41	2003-04 figure was inaccurate; current figure is from the UW Office of Institutional Studies.
WASHINGTON STATE	
	Includes WSU Vancouver, WSU Tri-Cities, WSU Energy Library, WSU Spokane, and WSU Intercollegiate College of Nursing. Volumes held June 30, 2004 revised to 2,197,160.
1.b.i	Includes e-books for the first time.
4	Decrease due to serial cancellation project.
21	Figure reflects one-time expenditures only. Excludes monographs.
23.a	Includes large authority control project (OCLC) costs.
24	Received a one-time allocation for computer hardware.
26.a	Three FTE positions abolished on the Main Campus for general faculty salary increases.
34-35	Due to a consolidation of a new Access Services unit, data are unavailable until 2005-06.
WASHINGTON U.-ST. LOUIS	
	Includes Biology, Business, Chemistry, East Asian, Earth and Planetary Sciences, Mathematics, Music, Physics, and Social Work libraries.
4	Decrease due to 2003-04 count including some serials in aggregator databases. Also, some duplicate titles cancelled in print if title is received in electronic form.
15.c	Includes $693,569 expended for electronic monographs.

QUESTION NUMBER	NOTES
19	Includes renovation expenses for central library now included in the budget and amortized over 30 years; also includes other facilities expenses that are a part of library budget, both for a total of $6,134,003; and also includes fringes in the amount of $1,668,364.
21	Increase due to an improved ability to identify charges and report them correctly.
22	Increase due to an increased emphasis on electronic serials and a decrease of duplicate serials in paper format.
23.b	Includes Health Science Library only.
31	Data unavailable for Law Library.
35	2003-04 figure included renewals; current figure does not.
36	2003-04 figure revised to 160.
37	Corrected from previous reports; previous reports included MD fields whereas number now being reported is PhD fields only.
WATERLOO	
	All figures are as of April 30, 2005 and include University Map Library, Musagetes Architecture Library (opened September 2004), and the Optometry Learning Resource Centre.
4	Includes 7,017 print serials and 13,011 electronic.
15-25	Expenditures as reported in Canadian dollars: (15a) $1,098,133; (15b) $5,160,968; (15c) $1,221,486; (15d) $260,952; (15) $7,741,539; (16) $68,550; (17a) $2,515,917; (17b) $3,496,416; (17c) $746,059; (17) $6,758,392; (19) $791,053; (20) $15,359,534; (21) $1,221,486; (22) $3,885,065; (23a) $0; (23b) $0; (24) $216,998; (25) $86,679.
15.b	Includes expenditures for print and e-resources.
22	Includes expenditures for electronic monograph and serial titles included in publishers' packages.
WAYNE STATE	
	All figures are as of September 30, 2005 and include Reuther Archives of Labor and Urban Affairs.
1	Includes government documents.
4	Figure is taken from the online system, which does not currently provide a count of purchased vs. not-purchased serials.
9, 11	Figures are estimated.
15.c	Total cannot be disaggegated from monograph expenditures (15a) and serial expenditures (15b).
19	Includes fringe benefits.
21	2003-04 figure included a non-recurring, gift-funded expenditure of $100,000.
32, 33	Totals for 2002-03 and 2003-04 were incorrectly calculated (included reserve circulations). The correct totals are 2002-03: (32) 165,458; (33) 239,713; and 2003-04: (32) 169,328; (33) 243,948.
36	Includes Ed. D. degrees.
WESTERN ONTARIO	
	All figures are as of April 30, 2005 and include: Archives and Research Collections Centre, Business Library, Education Library, Music Library, Allyn & Betty Taylor Library (Engineering, Science, Medicine & Dentistry, Health Sciences) and The D.B. Weldon Library (Arts & Humanities, Social Science, and Information & Media Studies).
1.a	Includes 148,126 e-monographs.
1.b	Decrease due to more withdrawals during 2004-05.
2	Includes 7,938 e-monographs.

QUESTION NUMBER	NOTES
4.a	Increase due to e-journals not purchased, as well as system changes which make it difficult to obtain reliable automated counts.
8, 13	Decrease due to a transition year in which computer and video files could no longer be counted manually.
11	Includes one collection of 1.6 million photographic negatives.
15-25	Expenditures as reported in Canadian dollars: (15a) $2,209,475; (15b) $8,874,837; (15c) $0; (15d) $0; (15) $11,084,312; (16) $151,905; (17a) $3,991,201; (17b) $3,563,903; (17c) $500,161; (17) $8,055,265; (19) $755,098; (20) $20,046,580; (21) $714,794; (22) $6,428,684; (23a) $107,230; (23b) $0; (24) $347,550; (25) $0.
21	Includes the Law Library.
23.a, 24	Centrally funded and includes the Law Library.
35	Requests from other libraries are handled centrally and include the Law Library.
WISCONSIN	
	Includes General Library System: Memorial Library, Undergraduate, Music, Art, Business, Social Science, Geography, Social Work, Biology, Chemistry, Geology and Geophysics, Mathematics, Physics, Steenbock Agricultural and Life Sciences, Ebling Medical, Law, Wendt Engineering, School of Education, School of Library and Information Studies, Map, Primate, and Wisconsin Historical Society.
4.a	Increase reflects addition of Wisconsin State Historical Society serials not previously reported.
7	An inconsistency in reporting of data has been discovered; the procedure for reporting for 2005-06 will be re-evaluated.
9	Decrease due to a correction to previous reporting for Special Collections.
12	Increase due to the library receiving a significant gift collection.
23.a	Figure reflects an increase in fees for basic services and addition of one time project costs.
YALE	
	Includes all 22 libraries of the Yale University Library, including Law and Medical. Volumes held June 30, 2004 revised to 11,728,447 to include 2004-05 aggregated electronic book purchases, including: Ebrary, IT Pro, Making of Modern Law, Evans Digital Edition, Early English Books Online, and Making of Modern Economy, 18th Century Collections Online, and EEBO (these last two were acquired, but not reported, in 2003-04).
8	Figure reflects only those files added; a cumulative figure is not reported. Computer files added in 2003-04 is 945.
33	Decreased count is largely result of redefining circulation; previous years' counts did not extract internal tracking circulation charges.
YORK	
	All figures are as of April 30, 2005 and include Scott Library (which houses Sound & Moving Image, Maps, Archives & Special Collections); Peter F. Bronfman Business Library; Steacie Science and Engineering Library; Osgoode Hall Law School Library; Leslie Frost Library on Glendon Campus.
12	Decrease due to a weeding of collection.
15-25	Expenditures as reported in Canadian dollars: (15a) $2,937,027; (15b) $6,043,553; (15c) $500,000; (15d) $0; (15) $9,480,580; (16) $180,558; (17a) $4,769,852; (17b) $4,767,011; (17c) $1,049,524; (17) $10,586,387; (19) $4,779,120; (20) $25,026,645; (21) $168,510; (22) $3,760,067; (23a) $69,776; (23b) $0; (24) $307,601; (25) $7,042.
19	Decrease due to 2003-04 figure including renovation costs.

APPENDIX A – ARL MEMBER LIBRARIES AS OF JANUARY 1, 2005

The Association of Research Libraries (ARL) represents the interests of 123 libraries that serve major North American research institutions. The ARL Statistics and Measurement program is organized around identifying, collecting, analyzing, and distributing quantifiable information describing the characteristics of research libraries.

Institution	Category	Full Name of Institution	Location
Alabama	S	University of Alabama	Tuscaloosa, Alabama
Alberta	C	University of Alberta	Edmonton, Alberta
Arizona	S	University of Arizona	Tucson, Arizona
Arizona State	S	Arizona State University	Tempe, Arizona
Auburn	S	Auburn University	Auburn, Alabama
Boston	P	Boston University	Boston, Massachusetts
Boston College	P	Boston College	Boston, Massachusetts
Brigham Young	P	Brigham Young University	Provo, Utah
British Columbia	C	University of British Columbia	Vancouver, British Columbia
Brown	P	Brown University	Providence, Rhode Island
Berkeley, California	S	University of California, Berkeley	California, Berkeley
California, Davis	S	University of California, Davis	Davis, California
California, Irvine	S	University of California, Irvine	Irvine, California
California, Los Angeles	S	University of California, Los Angeles	Los Angeles, California
California, Riverside	S	University of California, Riverside	Riverside, California
California, San Diego	S	University of California, San Diego	La Jolla, California
California, Santa Barbara	S	University of California, Santa Barbara	Santa Barbara, California
Case Western Reserve	P	Case Western Reserve University	Cleveland, Ohio
Chicago	P	University of Chicago	Chicago, Illinois
Cincinnati	S	University of Cincinnati	Cincinnati, Ohio
Colorado	S	University of Colorado	Boulder, Colorado
Colorado State	S	Colorado State University	Fort Collins, Colorado
Columbia	P	Columbia University	New York, New York
Connecticut	S	University of Connecticut	Storrs, Connecticut
Cornell	P	Cornell University	Ithaca, New York
Dartmouth	P	Dartmouth College	Hanover, New Hampshire
Delaware	S	University of Delaware	Newark, Delaware
Duke	P	Duke University	Durham, North Carolina
Emory	P	Emory University	Atlanta, Georgia
Florida	S	University of Florida	Gainesville, Florida
Flordia State	S	Florida State University	Tallahassee, Florida
George Washington	P	George Washington University	Washington, D.C.
Georgetown	P	Georgetown University	Washington, D.C.
Georgia	S	University of Georgia	Athens, Georgia
Georgia Tech	S	Georgia Institute of Technology	Atlanta, Georgia
Guelph	C	University of Guelph	Guelph, Ontario
Harvard	P	Harvard University	Cambridge, Massachusetts
Hawaii	S	University of Hawaii	Honolulu, Hawaii
Houston	S	University of Houston	Houston, Texas
Howard	P	Howard University	Washington, D.C.
Illinois, Chicago	S	University of Illinois at Chicago	Chicago, Illinois
Illinois, Urbana	S	University of Illinois at Urbana	Urbana, Illinois
Indiana	S	Indiana University	Bloomington, Indiana
Iowa	S	University of Iowa	Iowa City, Iowa
Iowa State	S	Iowa State University	Ames, Iowa
Johns Hopkins	P	Johns Hopkins University	Baltimore, Maryland
Kansas	S	University of Kansas	Lawrence, Kansas
Kent State	S	Kent State University	Kent, Ohio
Kentucky	S	University of Kentucky	Lexington, Kentucky
Laval	C	Laval University	Quebec, Quebec
Louisiana State	S	Louisiana State University	Baton Rouge, Louisiana
Louisville	S	University of Louisville	Louisville, Kentucky
McGill	C	McGill University	Montreal, Quebec
McMaster	C	McMaster University	Hamilton, Ontario
Manitoba	C	University of Manitoba	Winnipeg, Manitoba
Maryland	S	University of Maryland	College Park, Maryland
Massachusetts	S	University of Massachusetts	Amherst, Massachusetts
MIT	P	Massachusetts Institute of Technology	Cambridge, Massachusetts
Miami	P	University of Miami	Coral Gables, Florida
Michigan	S	University of Michigan	Ann Arbor, Michigan
Michigan State	S	Michigan State University	East Lansing, Michigan
Minnesota	S	University of Minnesota	Minneapolis, Minnesota

S: U.S. public univ.; P: U.S. private univ.; C: Canadian univ.; N: U.S. non-univ.; X: Canadian non-univ.

Institution	Category	Full Name of Institution	Location
Missouri	S	University of Missouri	Columbia, Missouri
Montreal	C	University of Montreal	Montreal, Quebec
Nebraska	S	University of Nebraska-Lincoln	Lincoln, Nebraska
New Mexico	S	University of New Mexico	Albuquerque, New Mexico
New York	P	New York University	New York, New York
North Carolina	S	University of North Carolina	Chapel Hill, North Carolina
North Carolina State	S	North Carolina State University	Raleigh, North Carolina
Northwestern	P	Northwestern University	Evanston, Illinois
Notre Dame	P	University of Notre Dame	Notre Dame, Indiana
Ohio	S	Ohio University	Athens, Ohio
Ohio State	S	Ohio State University	Columbus, Ohio
Oklahoma	S	University of Oklahoma	Norman, Oklahoma
Oklahoma State	S	Oklahoma State University	Stillwater, Oklahoma
Oregon	S	University of Oregon	Eugene, Oregon
Pennsylvania	P	University of Pennsylvania	Philadelphia, Pennsylvania
Pennsylvania State	S	Pennsylvania State University	University Park, Pennsylvania
Pittsburgh	S	University of Pittsburgh	Pittsburgh, Pennsylvania
Princeton	P	Princeton University	Princeton, New Jersey
Purdue	S	Purdue University	West Lafayette, Indiana
Queen's	C	Queen's University	Kingston, Ontario
Rice	P	Rice University	Houston, Texas
Rochester	P	University of Rochester	Rochester, New York
Rutgers	S	Rutgers University	New Brunswick, New Jersey
Saskatchewan	C	University of Saskatchewan	Saskatoon, Saskatchewan
South Carolina	S	University of South Carolina	Columbia, South Carolina
Southern California	P	University of Southern California	Los Angeles, California
Southern Illinois	S	Southern Illinois University	Carbondale, Illinois
SUNY-Albany	S	University at Albany, State University of New York	Albany, New York
SUNY-Buffalo	S	University at Buffalo, State University of New York	Buffalo, New York
SUNY-Stony Brook	S	State University of New York at Stony Brook	Stony Brook, New York
Syracuse	P	Syracuse University	Syracuse, New York
Temple	S	Temple University	Philadelphia, Pennsylvania
Tennessee	S	University of Tennessee	Knoxville, Tennessee
Texas	S	University of Texas	Austin, Texas
Texas A&M	S	Texas A&M University	College Station, Texas
Texas Tech	S	Texas Tech University	Lubbock, Texas
Toronto	C	University of Toronto	Toronto, Ontario
Tulane	P	Tulane University	New Orleans, Louisiana
Utah	S	University of Utah	Salt Lake City, Utah
Vanderbilt	P	Vanderbilt University	Nashville, Tennessee
Virginia	S	University of Virginia	Charlottesville, Virginia
Virginia Tech	S	Virginia Polytechnic Institute & State University	Blacksburg, Virginia
Washington	S	University of Washington	Seattle, Washington
Washington State	S	Washington State University	Pullman, Washington
Washington U.-St. Louis	P	Washington University	St. Louis, Missouri
Waterloo	C	University of Waterloo	Waterloo, Ontario
Wayne State	S	Wayne State University	Detroit, Michigan
Western Ontario	C	University of Western Ontario	London, Ontario
Wisconsin	S	University of Wisconsin	Madison, Wisconsin
Yale	P	Yale University	New Haven, Connecticut
York	C	York University	North York, Ontario
Boston Public Library	N	Boston Public Library	Boston, Massachusetts
Canada Inst. SciTech Info.	X	Canada Inst. for Scientific & Technical Information	Ottawa, Ontario
Center for Research Libs.	N	Center for Research Libraries	Chicago, Illinois
Library of Congress	N	Library of Congress	Washington, D.C.
Natl. Agricultural Lib.	N	National Agricultural Library	Beltsville, Maryland
Lib. & Archives of Canada	X	The Library and Archives of Canada	Ottawa, Ontario
Natl. Library of Medicine	N	National Library of Medicine	Bethesda, Maryland
New York Public Library	N	New York Public Library	New York, New York
New York State Library	N	New York State Library	Albany, New York
Smithsonian Institution	N	Smithsonian Institution	Washington, D.C.

S: U.S. public univ.; P: U.S. private univ.; C: Canadian univ.; N: U.S. non-univ.; X: Canadian non-univ.

APPENDIX B – BIBLIOGRAPHY

PART A: Selected Articles from the ARL Newsletter

"The ARL Membership Criteria Index." *ARL: A Bimonthly Newsletter of Research Library Issues and Actions* 197 (April 1998): 9.

"Assessing ILL/DD Services: New Cost-Effective Alternatives." *ARL: A Bimonthly Report on Research Library Issues and Actions from ARL, CNI, and SPARC* 236 (October 2004): 9.

Askew Waller, Consuella, and Kaylyn Hipps. "Using LibQUAL+ and Developing a Culture of Assessment in Libraries." *ARL: A Bimonthly Report on Research Library Issues and Actions from ARL, CNI, and SPARC* 221 (April 2002): 10-11.

Case, Mary M. "A Snapshot in Time: ARL Libraries and Electronic Journal Resources." *ARL: A Bimonthly Newsletter of Research Library Issues and Actions* 235 (August 2004): 1-10.

_______. "The Impact of Serial Costs on Library Collections." *ARL: A Bimonthly Report on Research Library Issues and Actions from ARL, CNI, and SPARC* 218 (October 2001): 9.

Case, Mary M. and Judith Matz. "Framing the Issue: Open Access." *ARL: A Bimonthly Report on Research Library Issues and Actions from ARL, CNI, and SPARC* 226 (February 2003): 8-10.

Case, Mary M. and Prudence Adler. "Promoting Open Access." *ARL: A Bimonthly Report on Research Library Issues and Actions from ARL, CNI, and SPARC* 220 (February 2002): 1-5.

"Collections & Access for the 21st-Century Scholar: Changing Roles of Research Libraries." Report from the ARL Collections & Access Issues Task Force. *ARL: A Bimonthly Report on Research Library Issues and Actions from ARL, CNI, and SPARC* 225 (December 2002).

Cook, Colleen, Fred Heath, and Bruce Thompson. "A Brief LibQUAL+ Phase One Progress Report." *ARL: A Bimonthly Report on Research Library Issues and Actions from ARL, CNI, and SPARC* 219 (December 2001): 7.

Crowe, William. "The End of History? Reflections on a Decade." *ARL: A Bimonthly Report on Research Library Issues and Actions from ARL, CNI, and SPARC* 226 (February 2003): 12-13.

Crow, Raym. "The Case for Institutional Repositories: A SPARC Position Paper." *ARL: A Bimonthly Report on Research Library Issues and Actions from ARL, CNI, and SPARC* 223 (August 2002): 1-4.

"The Future of Human Resources in Canadian Libraries." *ARL*, no. 240 (June 2005): 9.

Grosetta Nardini, Holly. "Building a Culture of Assessment." *ARL: A Bimonthly Report on Research Library Issues and Actions from ARL, CNI, and SPARC* 218 (October 2001): 11.

Groves, Richard. "Sharing Best Practices by Disseminating Assessment Results via the Web." *ARL: A Bimonthly Report on Research Library Issues and Actions from ARL, CNI, and SPARC* 236 (October 2004): 6.

Guedon, Jean-Claude. "Beyond Core Journals and Licenses: The Paths to Reform Scientific Publishing." *ARL: A Bimonthly Report on Research Library Issues and Actions from ARL, CNI, and SPARC* 218 (October 2001): 1-8.

Hahn, Karla. "The State of the Large Publisher Bundle: Findings from an ARL Member Survey." *ARL: A Bimonthly Report on Research Library Issues and Actions from ARL, CNI, and SPARC* 245 (April 2006).

Hipps, Kaylyn. "Diversity in the U.S. ARL Library Workforce." *ARL: A Bimonthly Report on Research Library Issues and Actions from ARL, CNI, and SPARC* 246 (June 2006): 1-2.

Johnson, Richard K. "Whither Competition?" *ARL: A Bimonthly Report on Research Library Issues and Actions from ARL, CNI, and SPARC* 217 (August 2001): 12-14.

Kyrillidou, Martha. "The Future of Librarians in the U.S. Workforce." *ARL: A Bimonthly Report on Research Library Issues and Actions from ARL, CNI, and SPARC* 246 (June 2006): 5.

Kyrillidou, Martha. "ARL University Libraries' Spending Trends." *ARL: A Bimonthly Report on Research Library Issues and Actions from ARL, CNI, and SPARC* 242 (October 2005): 10.

Kyrillidou, Martha. "LibQUAL+™ in 2004." *ARL: A Bimonthly Report on Research Library Issues and Actions from ARL, CNI, and SPARC* 236 (October 2004): 6.

______. "Serials Trends Reflected in the ARL Statistics 2002-03." *ARL: A Bimonthly Report on Research Library Issues and Actions from ARL, CNI, and SPARC* 234 (June 2004): 14-15.

Kyrillidou, Martha, and William Crowe. "In search of new measures." *ARL: A Bimonthly Newsletter of Research and Library Issues and Actions* 197 (April 1998): 8-10.

Kyrillidou, Martha, and Kaylyn Hipps. "Symposium on Measuring Library Service Quality." *ARL: A Bimonthly Report on Research Library Issues and Actions from ARL, CNI, and SPARC* 215 (April 2001): 9-11.

Lippincott, Sarah, and Martha Kyrillidou. "How ARL University Communities Access Information: Highlights From LibQUAL+™." *ARL: A Bimonthly Newsletter of Research Library Issues and Actions* 236 (October 2004): 7-8.

Lynch, Clifford. "Institutional Repositories: Essential Infrastructure for Scholarship in the Digital Age." *ARL: A Bimonthly Report on Research Library Issues and Actions from ARL, CNI, and SPARC* 226 (February 2003): 1-7.

"Making Library Assessment Work." *ARL: A Bimonthly Report on Research Library Issues and Actions from ARL, CNI, and SPARC* 240 (June 2005): 9.

Shim, Wonsik "Jeff", Charles McClure, and John Carlo Bertot. "Measures and Statistics for Research Library Networked Services: ARL E-Metrics Phase II Report." *ARL: A Bimonthly Report on Research Library Issues and Actions from ARL, CNI, and SPARC* 219 (December 2001): 8-9.

Stuber, Peter. "Where does the Free Online Scholarship Movement Stand Today?" *ARL: A Bimonthly Report on Research Library Issues and Actions from ARL, CNI, and SPARC* 220 (February 2002): 9-15.

Unsworth, John M. "The Crisis in Scholarly Publishing in the Humanities." *ARL: A Bimonthly Report on Research Library Issues and Actions from ARL, CNI, and SPARC* 228 (June 2003): 1-4.

Wetzel, Karen, and Mary Jackson. "Portal Functionality Provided by ARL Libraries: Results of an ARL Survey." *ARL: A Bimonthly Report on Research Library Issues and Actions from ARL, CNI, and SPARC* 222 (June 2002): 7-9.

Wilder, Stanley. "New Hires in Research Libraries: Demographic Trends and Hiring Priorities." *ARL: A Bimonthly Report on Research Library Issues and Actions from ARL, CNI, and SPARC* 221 (April 2002): 5-8.

Young, Mark. "*ARL Salary Survey* Highlights." *ARL: A Bimonthly Report on Research Library Issues and Actions from ARL, CNI, and SPARC* 246 (June 2006): 4.

PART B: Related Books and Articles Published in Other Sources

Arms, William. "Quality Control in Scholarly Publishing on the Web." *The Journal of Electronic Publishing* 8, no. 1 (August 2002).

Baumol, W. J., and M. Marcus. *Economics of Academic Libraries*. Washington, D.C.: American Council of Education, 1973.

Bertot, John Carlo, Charles McClure, and Joe Ryan. *Statistics and Performance Measures for Public Library Networked Services*. Chicago: American Library Association, 2001.

Blixrud, Julia. "The Association of Research Libraries Statistics and Measurement Program: From Descriptive Data to Performance Measures." *Proceedings from the 4th Northumbria International Conference on Performance Measurement in Libraries and Information Services*, edited by Joan Stein, Martha Kyrillidou, and Denise Davis. Washington, D.C.: Association of Research Libraries, 2002.

Carlson, Scott. "The Deserted Library: As Students Work Online, Reading Rooms Empty Out – Leading Some Campuses to Add Starbucks." The Chronicle of Higher Education, November 16, 2001.

Clapp, V. W. *The Future of the Research Library*. Urbana: University of Illinois Press, 1964.

Coffman, Steve. "Building Earth's Largest Library: Driving into the Future." *Searcher* 7, no. 3 (March 1999).

Competition Commission of the United Kingdom. *Reed Elsevier plc and Harcourt General, Inc.: A report on the proposed merger*, presented to Parliament by the Secretary of State and Trade and Industry by Command of Her Majesty, July 2001.

Cook, Colleen, Fred Heath, Martha Kyrillidou, and Duane Webster. "The Forging of Consensus: A Methodological Approach to Service Quality Assessment in Research Libraries – the LibQUAL+ Experience." *Proceedings from the 4th Northumbria International Conference on Performance Measurement in Libraries and Information Services*, edited by Joan Stein, Martha Kyrillidou, and Denise Davis. Washington, D.C.: Association of Research Libraries, 2002.

Cook, Colleen, Fred Heath, and Bruce Thompson. "Score norms for improving library service quality: A LibQUAL+ study." *portal: Libraries and the Academy* 2 (2002): 13-26.

Cook, Colleen, Fred Heath, Bruce Thompson, and Duane Webster. "LibQUAL+™ Preliminary Results from 2002." *Performance Measurement and Metrics* 4, no. 1 (2003): 38-47.

Cook, Colleen, Fred Heath, Bruce Thompson, and Russell Thompson. *The Search for New Measures: The ARL LibQUAL+Project--A Preliminary Report.* Baltimore: Johns Hopkins University Press, 2001. Accessed on the Internet at <http://muse.jhu.edu/demo/pla/1.1cook.html> on August 22, 2001.

_______. "LibQUAL+: Service Quality Assessment in Research Libraries." *IFLA Journal* 27 (2001): 264-8.

Council on Library and Information Resources. *A Different Approach to the Evaluation of Research Libraries.* Research Brief 6. Washington, D.C.: The Council, 1998.

Crow, Raym. "The Case for Institutional Repositories: A SPARC Position Paper." Washington, DC: Scholarly Publishing & Academic Resources Coalition, 2002.

Cummings, Anthony M., et al. *University Libraries and Scholarly Communication: A Study Prepared for The Andrew W. Mellon Foundation.* Washington, D.C.: Association of Research Libraries, 1992.

Cummings, M. M. *The Economics of Research Libraries.* Washington, D.C.: Council on Library Resources, 1986.

Downs, Robert B. "The Growth of Research Collections." *Library Trends: American Library History: 1876-1976* 25 (July 1976): 55-80.

_______. "Uniform Statistics for Library Holdings." *Library Quarterly* 16 (January 1946): 63-69.

Drake, Miriam A. "Forecasting Academic Library Growth." *College and Research Libraries* 37 (January 1976): 53-59.

Frazier, Kenneth. "The Librarians' Dilemma: Contemplating the Costs of the 'Big Deal,'" *D-Lib Magazine* 7, no. 3 (March 2001), viewed online at http://www.dlib.org/dlib/march01/frazier/03frazier.html.

Gooden, Paul, Matthew Owen, and Sarah Simon. *Scientific Publishing: Knowledge is Power.* New York: Morgan Stanley, 2002.

Hayes, Robert M. *Strategic Management for Academic Libraries: A Handbook.* Westport, Conn.: Greenwood Press, 1993.

Heath, Fred, Colleen Cook, Martha Kyrillidou, and Bruce Thompson. "ARL Index and other validity correlates of LibQUAL+ scores." *portal: Libraries and the Academy* 2 (2002): 27-42.

Hernon, Peter, and Robert E. Dugan. *Action Plan for Outcomes Assessment in Your Library.* Chicago: American Library Association, 2002.

Hernon, Peter, and John R. Whitman. *Delivering Satisfaction and Service Quality.* Chicago: American Library Association , 2001.

Hernon, Peter, and Ellen Altman. *Assessing Service Quality.* Chicago: American Library Association, 1998.

Hightower, Christy, Julie Sih, and Adam Tilgham. "Recommendations for Benchmarking Web Site Usage Among Academic Libraries." *College and Research Libraries* 59 (January 1998): 61-79.

Steve Hiller, Martha Kyrillidou and Jim Self, "Assessment in North American research libraries: a preliminary report card." *Performance Measurement and Metrics* 7 (2) (2006): 100-6.

Jackson, Mary E. *Measuring the Performance of Interlibrary Loan Operations in North American Research & College Libraries.* Washington, D.C.: Association of Research of Libraries, 1998.

Johnson, Richard K. "A Question of Access: SPARC, BioOne, and Society-Driven Electronic Publishing." *D-Lib Magazine* (May 2000), accessed on the Internet at <http://www.dlib.org/dlib/may00/johnson/05johnson.html> on August 21, 2001.

Kling, Rob, Lisa Spector, and Geoff McKim. "The Guild Model." *The Journal of Electronic Publishing* 8, no. 1 (August 2002).

Kuhlman, A. F. "Two ARL Approaches to Counting Holdings of Research Libraries." *College and Research Libraries* 21 (May 1960): 207-11.

Kyrillidou, Martha. "Library Assessment as a collaborative enterprise." *Resource Sharing and Information Networks* 1/2 (Special 2005-06 issue on the theme "Creative Collaborations: Libraries Within Their Institutions and Beyond"): 73-87.

Kyrillidou, Martha. "From Input and Output Measures to Quality and Outcome Measures, or, from the User in the Life of the Library to the Library in the Life of the User." *The Journal of Academic Librarianship* 28, no. 1 (January-March 2002): 42-46.

_______. "To Describe and Measure the Performance of North American Research Libraries." *IFLA Journal* 27 (2001, No. 4): 257-63.

_______. "Research Library Trends: *ARL Statistics*." Washington, D.C.: Association of Research Libraries, 2001. Accessed on the Internet at <http://www.arl.org/stats/arlstat/jal99.html> on August 22, 2001.

Kyrillidou, Martha, and Fred Heath, eds. "Measuring Service Quality." *Library Trends* 49 (Spring 2001).

Lakos, Amos. "Culture of Assessment as a Catalyst for Organizational Culture Change in Libraries." *Proceedings from the 4th Northumbria International Conference on Performance Measurement in Libraries and Information Services*, edited by Joan Stein, Martha Kyrillidou, and Denise Davis. Washington, D.C.: Association of Research Libraries, 2002.

Lavoie, Brian, Lynn Silipigni Connaway, and Lorcan Demsey. "Anatomy of aggregate collections: the example of Google Print for libraries" *D-Lib Magazine* (September 2005) 11 (9): http://www.dlib.org/dlib/september05/lavoie/09lavoie.html .

Lynch, Clifford A., and Joan K. Lippincott. "Institutional repository deployment in the United States as of early 2005" *D-Lib Magazine* (September 2005) 11 (9): http://www.dlib.org/dlib/september05/lynch/09lynch.html .

Mekkawi, Mod. "The ARL Library Index as a Decision-Making Tool." *College and Research Libraries* 43 (September 1982): 396-401.

Miller, Rush, and Sherrie Schmidt. "E-Metrics: Measures for Electronic Resources." *Proceedings from the 4th Northumbria International Conference on Performance Measurement in Libraries and Information Services*, edited by Joan Stein, Martha Kyrillidou, and Denise Davis. Washington, D.C.: Association of Research Libraries, 2002.

Miller, Rush, Sherrie Schmidt, and Martha Kyrillidou. "New Initiatives in Performance Measures." *Global Issues in 21st Century Research Librarianship*, edited by Sigrun Klara Hannesdottir. Helsinki: NORDINFO, 2002.

Molyneux, Robert E. "What Did Rider Do? An Inquiry into the Methodology of Fremont Rider's *The Scholar and the Future of the Research Library*." *Libraries and Culture* 29 (Summer 1994): 297-325.

_______. "Patterns, Processes of Growth, and the Projection of Library Size: A Critical Review of the Literature on Academic Library Growth." *Library and Information Science Research* 8 (January-March 1986): 5-28.

_______. *The Gerould Statistics 1907/08 - 1961/62*. Washington, D.C.: Association of Research Libraries, 1986.

Molyneux, Robert E., and Robert V. Williams. "Measuring the Internet." *Annual Review of Information Science and Technology* 34 (1999-2000): 287-339.

Morris, Dilys E., Collin B. Hobert, Lori Osmus, and Gregory Wool. "Cataloging Staff Costs Revisited." *LRTS* 44 (2000): 70-83.

Nitecki, Danuta, and Toni Olshen. "Developing Service Quality Measurement Skills Through and Online Course for Librarians." *Proceedings from the 4th Northumbria International Conference on Performance Measurement in Libraries and Information Services*, edited by Joan Stein, Martha Kyrillidou, and Denise Davis. Washington, D.C.: Association of Research Libraries, 2002.

O'Connor, Lisa, Carolyn Radcliff, and Julie Gedeon. "Applying Systems Design and Item Response Theory to the Problem of Measuring Information Literacy Skills." *College and Research Libraries* 63, no. 6 (2002): 528-543.

Piternick, George. "ARL Statistics—Handle With Care." *College and Research Libraries* 38 (September 1977): 419-23.

_______. "Library Growth and Academic Quality." *College and Research Libraries* 24 (May 1963): 223-29.

Rider, Fremont. *The Scholar and the Future of the Research Library*. New York: Hadham Press, 1944.

Roche, Marilyn M. *ARL/RLG Interlibrary Loan Cost Study*. Washington, D.C.: Association of Research Libraries, 1993.

Rosenblatt, Susan. "Developing Performance Measures for Library Collections and Services." Chap. 18 in *The Mirage of Continuity: Reconfiguring Academic Information Resources for the 21st Century,* edited by Brian L. Hawkins and Patricia Battin. Washington, D.C.: Council on Library and Information Resources, 1998.

Shim, Wonsik, Charles McClure, and John Carlo Bertot. "Preliminary Statistics and Measures for ARL Libraries to Describe Electronic Resources and Services." *Proceedings from the 4th Northumbria International Conference on Performance Measurement in Libraries and Information Services,* edited by Joan Stein, Martha Kyrillidou, and Denise Davis. Washington, D.C.: Association of Research Libraries, 2002.

Stubbs, Kendon. "Access and ARL membership criteria." *Proceedings of the 125th Meeting of the Association of Research Libraries* (1993): 117-22.

_______. "University Libraries: Standards and Statistics." *College and Research Libraries* 42 (November 1981): 527-38.

_______. "Apples and Oranges and ARL Statistics." *Journal of Academic Librarianship* 14 (September 1988): 231-35.

_______. "On the ARL Library Index." Paper presented at the 108th meeting of Research Libraries: Measurement, Management, Marketing (May 1986): Minneapolis, MN.

_______. "Lies, damned lies, ... and ARL Statistics?" Paper presented at the 108th meeting of Research Libraries: Measurement, Management, Marketing (May 1986): Minneapolis, MN.

Thompson, Bruce, Colleen Cook, and Russell Thompson. "Reliability and structure of LibQUAL+ scores." *portal: Libraries and the Academy* 2 (2002): 3-12.

van Westrienen, Gerard, and Clifford A. Lynch. "Academic Institutional Repositories: deployment status in 13 nations as of mid 2005" *D-Lib Magazine* (September 2005) 11 (9): http://www.dlib.org/dlib/september05/westrienen/09westrienen.html .

Weiner, Sharon. "Library quality and impact: Is there a relationship between new measures and traditional measures?" *Journal of Academic Librarianship* 31 (2005): 432-7.

Wilder, Stanley J. *The Age Demographics of Academic Librarians: A Profession Apart. A Report Based on Data from the ARL Annual Salary Survey.* Washington, D.C.: Association of Research Libraries, 1995.

Williams, Robert V. "The Making of Statistics of National Scope on American Libraries, 1836-1986: Purposes, Problems, and Issues." *Libraries and Culture* 26 (Spring 1991): 464-85.

Wysocki, Bernard. "Scholarly Journals' Premiere Status is Diluted by Web." *Wall Street Journal,* May 23, 2005; Page A1.

Made in the USA
Las Vegas, NV
29 September 2026

55281240R00083